my revision notes

AQA GCSE (9–1)
PHYSICS

Nick England

Photos reproduced by permission of: **p.27**, Fatbob/Fotolia; **p.90**, Martin Dohrn/Science Photo Library; **p.98**, Isabelle Limbach/iStockphoto/Thinkstock; **p.119**, SSPL/Getty images

Although every effort has been made to ensure that website addresses are correct at time of going to press, Hodder Education cannot be held responsible for the content of any website mentioned. It is sometimes possible to find a relocated web page by typing in the address of the home page for a website in the URL window of your browser.

Orders: please contact Hachette UK Distribution, Hely Hutchinson Centre, Milton Road, Didcot, Oxfordshire, OX11 7HH. Telephone: +44 (0)1235 827827. Email education@hachette.co.uk Lines are open from 9 a.m. to 5 p.m., Monday to Friday. You can also order through our website: www.hoddereducation.co.uk

ISBN 978 1 4718 5141 4

© Nick England 2017

First published in 2017 by
Hodder Education
An Hachette UK Company,
Carmelite House,
50 Victoria Embankment
London EC4Y 0DZ

Impression number   5

Year   2023

Cover photo © Ian Knaggs/Alamy Stock Photo

Typeset in Bembo Std Regular 11/13 by Integra Software Services Pvt. Ltd., Pondicherry, India

Printed by CPI Group (UK) Ltd, Croydon CR0 4YY

A catalogue record for this title is available from the British Library.

Get the most from this book

Everyone has to decide his or her own revision strategy, but it is essential to review your work, learn it and test your understanding. These Revision Notes will help you to do that in a planned way, topic by topic. Use this book as the cornerstone of your revision and don't hesitate to write in it — personalise your notes and check your progress by ticking off each section as you revise.

Tick to track your progress

Use the revision planner on pages iv and v to plan your revision, topic by topic. Tick each box when you have:

- revised and understood a topic
- tested yourself
- practised the exam questions and gone online to check your answers and complete the quick quizzes.

You can also keep track of your revision by ticking off each topic heading in the book. You may find it helpful to add your own notes as you work through each topic.

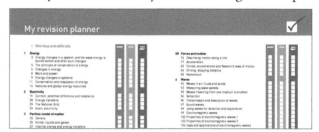

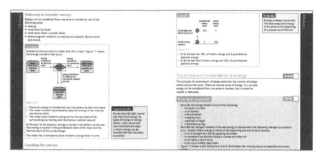

Features to help you succeed

Exam tips and summaries

Expert tips are given throughout the book to help you polish your exam technique in order to maximise your chances in the exam. The summaries provide a quick-check bullet list for each topic.

Typical mistakes

The author identifies the typical mistakes students make and explains how you can avoid them.

Now test yourself

These short, knowledge-based questions provide the first step in testing your learning. Answers are at the back of the book.

Definitions of key terms

Clear, concise definitions of essential key terms are provided where they first appear.

Exam practice

Practice exam questions are provided for each topic. Use them to consolidate your revision and practise your exam skills.

Online

Go online to check your answers to the exam practice questions and to try out the extra quick quizzes at **www.hoddereducation.co.uk/ myrevisionnotesdownloads**

Required practical

The exam board lists some practicals that you are required to do and understand. These are summarised under this heading.

Working scientifically

A section on this subject reminds you about the skills you have learnt throughout the course, and shows you how these are examined.

H Where this symbol appears, the text to the right of it relates to higher tier material.

My revision planner

	REVISED	TESTED	EXAM READY

Exam practice answers and quick quizzes at
www.hoddereducation.co.uk/myrevisionnotesdownloads

Countdown to my exams

6–8 weeks to go

- Start by looking at the specification — make sure you know exactly what material you need to revise and the style of the examination. Use the revision planner on pages iv and v to familiarise yourself with the topics.
- Organise your notes, making sure you have covered everything on the specification. The revision planner will help you to group your notes into topics.
- Work out a realistic revision plan that will allow you time for relaxation. Set aside days and times for all the subjects that you need to study, and stick to your timetable.
- Set yourself sensible targets. Break your revision down into focused sessions of around 40 minutes, divided by breaks. These Revision Notes organise the basic facts into short, memorable sections to make revising easier.

REVISED ☐

2–6 weeks to go

- Read through the relevant sections of this book and refer to the exam tips, summaries, typical mistakes and key terms. Tick off the topics as you feel confident about them. Highlight those topics you find difficult and look at them again in detail.
- Test your understanding of each topic by working through the 'Now test yourself' questions in the book. Look up the answers at the back of the book.
- Make a note of any problem areas as you revise, and ask your teacher to go over these in class.
- Look at past papers. They are one of the best ways to revise and practise your exam skills. Write or prepare planned answers to the exam practice questions provided in this book. Check your answers online and try out the extra quick quizzes at **www.hoddereducation.co.uk/ myrevisionnotesdownloads**
- Try out different revision methods. For example, you can make notes using mind maps, spider diagrams or flash cards.
- Track your progress using the revision planner and give yourself a reward when you have achieved your target.

REVISED ☐

One week to go

- Try to fit in at least one more timed practice of an entire past paper and seek feedback from your teacher, comparing your work closely with the mark scheme.
- Check the revision planner to make sure you haven't missed out any topics. Brush up on any areas of difficulty by talking them over with a friend or getting help from your teacher.
- Attend any revision classes put on by your teacher. Remember, he or she is an expert at preparing people for examinations.

REVISED ☐

The day before the examination

- Flick through these Revision Notes for useful reminders, for example the exam tips, summaries, typical mistakes and key terms.
- Check the time and place of your examination.
- Make sure you have everything you need — extra pens and pencils, tissues, a watch, bottled water, sweets.
- Allow some time to relax and have an early night to ensure you are fresh and alert for the examinations.

REVISED ☐

My exams

GCSE Physics Paper 1 (Topics 1–4)

Date:..

Time:..

Location: ....

GCSE Physics Paper 2 (Topics 5–8)

Date:..

Time:..

Location: ....

Working scientifically

Assessment objectives

When you study a science subject for GCSE, you gain a lot of knowledge and understanding of scientific ideas, techniques and procedures. However, good scientists do not just know facts, they apply their knowledge and understanding to solve unfamiliar problems. Scientists also have the skills to analyse and interpret information that they have gained from experimental observation, and then to draw conclusions. When GCSE examiners set your exam papers, they have these skills in mind.

You will have acquired these skills as your teacher led you through the course, but you also need to revise these skills as you work through this book. You must be **active** as you revise: reading the text will reinforce your knowledge and understanding, but you will only learn to apply your knowledge and understanding by **answering questions**. You will find that the questions in the book cover the practical work that you did in the course, including the ten required practicals.

The sections below offer a reminder of how you learnt to think as a scientist throughout your course.

Development of scientific thinking

1 Understand how scientific methods and theories develop over time. For example, the nuclear model of the atom was proposed after new evidence.
2 Use a variety of models to solve problems and make predictions. For example, we use the particle model of gases to explain how a gas exerts a pressure.
3 Consider the limitations of science and consider ethical issues. For example, the Big Bang Theory leaves questions unanswered; we should be aware of the ethical issues of generating electricity from fossil fuels.
4 Explain everyday and technological applications of science. For example, the knowledge of physics is used extensively in the design of cars.
5 Evaluate risks in a social context. For example, we must evaluate the risks and benefits of using radioactive sources in medicine.
6 Recognise the importance of peer review of results and of communicating results to a range of audiences. For example, we accept Newton's laws of motion as universally true, because they have been tested frequently and found to be true – and you have tested them too during your course.

Experimental skills and strategies

You have been taught about experimental skills throughout the course, and some of these ideas will be tested in your GCSE exams:
1 Develop hypotheses using scientific theories and explanations.
2 Plan experiments. For example, plan an experiment to measure the speed of sound.
3 Choose appropriate instruments, apparatus and techniques.
4 Carry out experiments appropriately and safely.
5 Recognise how many measurements should be taken to ensure a representative sample.
6 Make and record observations.
7 Evaluate your method and suggest possible improvements.

Analysis and evaluation

1 Present data using an appropriate method. For example, you need to be able to construct tables and plot graphs.
2 Translate data from one form to another. For example, you need to be able to interpret data from graphs.
3 Carry out mathematical analysis – there is plenty of practice in this book. For example:
   ○ Use an appropriate number of significant figures in calculations.
   ○ Find an arithmetic mean.

○ Change the subject of an equation. (**This is a vital skill.**)

○ Substitute numerical values into equations and use appropriate units for physical quantities.

○ Determine the slope of a linear graph.

○ Draw and use the slope of a tangent to a curve as a measure of the rate of change.

○ Understand the significance of the area under a graph and measure it by counting squares. For example, the area under a velocity–time graph is the distance travelled.

4 Make estimates of uncertainty.

5 Identify patterns and trends in data. Draw conclusions from given observations.

6 Comment on the extent to which data supports a hypothesis.

7 Evaluate data in terms of accuracy, precision, repeatability and reproducibility, and identify sources of systematic and random errors.

○ An accurate measurement is one that is close to the true value.

○ Measurements are precise if they cluster closely.

○ Measurements are repeatable when measurements, under similar circumstances, give similar results.

○ Random errors are caused by results varying in unpredictable ways, and can be reduced by making more measurements and reporting a mean value.

○ A systematic error is due to measurement results differing from the true value by a consistent amount each time. For example, an ammeter might always read too high a value of the current.

○ You should be able to recognise and ignore anomalous results.

8 Present coherent and logical accounts using experimental skills and analysis and evaluation.

Scientific, vocabulary, quantities, units, symbols and nomenclature

1 Use scientific vocabulary, terminology and definitions. For example, when you describe waves, the words *wavelength*, *amplitude* and *frequency* have precise meanings.

2 Recognise the importance of scientific quantities. For example, the gravitational field strength, $g = 9.8\,\text{N/kg}$.

3 Use the correct units. For example, joules, J; millimetres, mm.

4 Use prefixes and powers of ten for orders of magnitude.

○ tera	10^{12}		○ centi	10^{-2}
○ giga	10^{9}		○ milli	10^{-3}
○ mega	10^{6}		○ micro	10^{-6}
○ kilo	10^{3}		○ nano	10^{-9}

5 Interconvert units. For example 1000 milliseconds = 1 second.

6 Work to an appropriate number of significant figures. If data is given to 2 significant figures, give your answer to 2 significant figures.

1 Energy

Energy changes in a system, and the ways energy is stored before and after such changes

REVISED

A system is an object or group of objects. There are changes in the way energy is stored when a system changes.

- **Throwing an object upwards**
  When you throw a ball upwards, just after the ball leaves your hand, it has a store of kinetic energy. By the time the ball reaches its highest point, the kinetic energy has been transferred to a gravitational potential energy store. Just before you catch the ball, it has a store of kinetic energy again.

- **A moving object hitting an obstacle**
  If you drop a lump of clay, it sticks to the ground. Just before the clay hits the ground, it has a store of kinetic energy. After it has hit the ground, the kinetic energy has all been transferred to thermal energy, which is stored in the clay or in the surroundings. We hear a sound when the clay hits the ground, but the sound is quickly dissipated as thermal energy in the surroundings.

- **Accelerating a car with a constant force**
  A force on the car does work to accelerate the car. There is a store of chemical energy in the car's petrol. As the petrol burns, it transfers energy into the kinetic energy store of the moving car and also into the thermal store of the surroundings.

- **Heating a resistor**
  When a current flows through a resistor, energy is transferred by electrical work. Energy is transferred from the chemical store in the battery into the thermal store of the resistor and then into the thermal store of the surroundings.

> **Exam tip**
> Energy can be transferred from one store to another.

Energy stores

When a system changes, energy can be transferred from one energy store to another.

You will meet the following energy stores in your GCSE course:
- kinetic
- chemical
- gravitational potential
- magnetic
- elastic potential
- electrostatic
- thermal (or internal)
- nuclear.

> **Exam tip**
> Remember there are eight stores of energy.

Pathways to transfer energy

Energy can be transferred from one store to another by one of the following paths:

- heating
- work done by forces
- work done when a current flows
- electromagnetic radiation or mechanical radiation (shock waves and sound).

A battery drives an electric motor that lifts a mass. Figure 1.1 shows the energy transfers that occur.

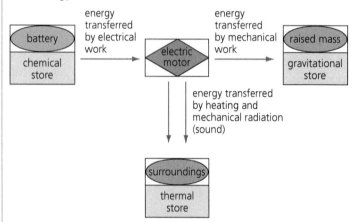

Figure 1.1

- Chemical energy is transferred from the battery by electrical work.
- The motor transfers gravitational potential energy to the mass by mechanical work.
- The motor also transfers energy to the thermal store of the surroundings by heating and mechanical radiation (sound).

At the start of the process, energy is stored in the battery; at the end, that energy is stored in the gravitational store of the mass and the thermal store of the surroundings.

The motor has a temporary store of kinetic energy when it turns.

Typical mistake

Do not describe light, sound and 'electrical energy' as types of energy or energy stores. Light, sound and electrical work are ways in which energy can be transferred from one store to another.

Counting the energy

Energy is a quantity that is measured in joules, J. Large quantities of energy are measured in kilojoules, kJ and megajoules, MJ.

$1\,kJ = 10^3\,J$

$1\,MJ = 10^6\,J$

The reason that energy is so important to us is that there is always the same energy at the end of a process as there was at the beginning.

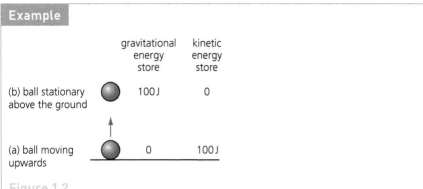

(b) ball stationary above the ground — gravitational energy store 100 J — kinetic energy store 0

(a) ball moving upwards — gravitational energy store 0 — kinetic energy store 100 J

Figure 1.2

Exam tip

Energy is always conserved. The total amount of energy is the same at the beginning of a process as at the end.

- At (a) the ball has 100 J of kinetic energy and 0 gravitational potential energy.
- At (b) the ball has 0 kinetic energy and 100 J of gravitational potential energy.

The principle of conservation of energy

REVISED

The principle of conservation of energy states that the amount of energy always remains the same. There are various stores of energy. In a process, energy can be transferred from one store to another, but it cannot be created or destroyed.

Now test yourself

TESTED

1 Describe the energy stored in each of the following:
  (a) hot water in a bath
  (b) a car battery
  (c) a litre of petrol
  (d) a moving train
  (e) a golf ball in flight
  (f) a stretched spring.
2 Describe the changes involved in the way energy is stored when the following changes to a system occur. Explain where energy is stored at the beginning and end of each process.
  (a) A car is brought to a halt by applying its brakes.
  (b) An aeroplane accelerates along a runway and takes off.
  (c) A cell lights a torch lamp.
  (d) A hot cup of coffee cools down.
3 Figure 1.3 shows a ball falling from A to B. Write down the missing values of potential and kinetic energy.

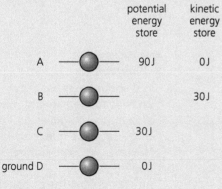

	potential energy store	kinetic energy store
A	90 J	0 J
B		30 J
C	30 J	
ground D	0 J	

Figure 1.3

Answers on p. 141

Changes in energy

Kinetic energy

The **kinetic energy** of a moving object can be calculated using the equation:

kinetic energy = 0.5 × mass × speed²

$$E_k = \frac{1}{2}mv^2$$

> kinetic energy, E_k, in joules, J
>
> mass, m, in kilograms, kg
>
> speed, v, in metres per second, m/s

Elastic potential energy

The amount of **elastic potential energy** stored in a stretched spring can be calculated using the equation:

elastic potential energy = 0.5 × spring constant × extension²

$$E_e = \frac{1}{2}ke^2$$

(assuming the limit of proportionality has not been exceeded).

> elastic potential energy, E_e, in joules, J
>
> spring constant, k, in newtons per metre, N/m
>
> extension, e, in metres, m

Gravitational potential energy

The **gravitational potential energy** gained by an object raised above the ground can be calculated using the equation:

gpe = mass × gravitational field strength × height

$$E_p = mgh$$

> gravitational potential energy, E_p, in joules, J
>
> gravitational field strength, g, in newtons per kilogram, N/kg
>
> height, h, in metres, m

Examples

1 Calculate the gravitational potential energy gained by a boy of mass 55 kg who climbs a flight of stairs 3.8 m high.
2 Calculate the elastic potential energy stored in a spring extended by 6.0 cm, which has a spring constant of 50 N/m.
3 Calculate the change in kinetic energy when a car of mass 1400 kg slows from 20 m/s to 10 m/s.

Answers

1 $E_p = mgh$

$\quad = 55 \times 9.8 \times 3.8$

$\quad = 2048\,J \approx 2000\,J$

2 $E_e = \frac{1}{2}ke^2$

$\quad = \frac{1}{2} \times 50 \times (0.06)^2$

$\quad = 0.09\,J$

3 $E_k = \frac{1}{2}mv_1^2 - \frac{1}{2}mv_2^2$

$\quad = \frac{1}{2} \times 1400 \times 20^2 - \frac{1}{2} \times 1400 \times 10^2$

$\quad = (700 \times 400) - (700 \times 100)$

$\quad = 210\,000\,J$ or 210 kJ

Typical mistake

You must always convert cm to m to get the energy in joules, J.

Exam tip

You should express your answer to the same number of significant figures as the data in the question. However, you will not be penalised over significant figures unless the question specifically asks for them.

Energy transfers and calculations

The equations for kinetic energy, elastic potential energy and gravitational potential energy may also be used to make calculations and predictions when a process causes energy to be transferred from one energy store to another.

Gravitational potential and kinetic energy stores

Example

A ball of mass 0.15 kg is thrown vertically upwards with a speed of 20 m/s.
1 Calculate the kinetic energy of the ball.
2 Calculate the maximum height to which the ball rises.

Answer

1 $E_k = \frac{1}{2}mv^2$

$\frac{1}{2} \times 0.15 \times 20^2$

$= 30\,J$

2 The energy in the ball's kinetic store is all transferred to the ball's gravitational potential energy store when it reaches its maximum height.
So the maximum E_p is 30 J.

$E_p = mgh$

$30 = 0.15 \times 9.8 \times h$

$h = \dfrac{30}{0.15 \times 9.8}$

$= 20.4\,m \approx 20\,m$

Maths note

The problem with the ball can be solved more quickly as follows:

E_k is transferred to E_p.

So,

$\frac{1}{2}mv^2 = mgh$

$h = \dfrac{v^2}{2g}$

$= \dfrac{20^2}{2 \times 9.8}$

$= 20.4\,m \approx 20\,m$

Now test yourself

TESTED

4 Calculate the increase in the gravitational energy store of a girl of mass 45 kg who climbs to the top of the Shard, which has a height of 310 m. Gravitational field strength $g = 9.8\,N/kg$.
5 Calculate the kinetic energy of a bullet with a mass of 20 g, travelling at a speed of 700 m/s.
6 A car with a mass of 900 kg increases its speed from 12 m/s to 18 m/s. Calculate the increase in the car's kinetic energy store. Express your answer in kilojoules.
7 A car suspension spring has a spring constant of 1800 N/m. Calculate the elastic potential energy stored in the spring when it is compressed by 10 cm.
8 A stone of mass 0.08 kg is dropped from a bridge into a river 12 m below.
   (a) Calculate the stone's gravitational potential energy 12 m above the river.
   (b) Calculate the speed of the stone as it hits the river.
   (c) The stone comes to rest on the riverbed. Explain where the stone's original gravitational potential energy is stored now.

9 Figure 1.4 shows a gymnast bouncing on a trampoline. She has a mass of 52 kg. After dropping from a height of 4.0 m, the trampoline has been stretched by 0.9 m.

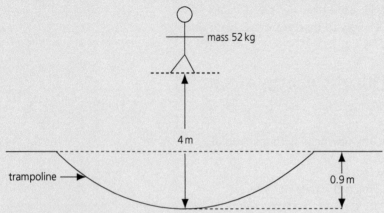

mass 52 kg

4 m

trampoline

0.9 m

Figure 1.4

(a) Calculate the gymnast's potential energy, 4.0 m above her lowest point.
(b) State the elastic potential energy stored in the trampoline after it has been stretched by 0.9 m.
(c) Use your answer to (b) to calculate the spring constant for the trampoline.

10 A bow has a spring constant of 300 N/m; it is used to shoot an arrow of mass 30 g. The bow string is pulled back 0.9 m before the arrow is released.
(a) Calculate the elastic potential energy stored in the bow when the string has been pulled back by 0.9 m.
(b) Assuming all the bow's elastic potential energy is transferred to the arrow's kinetic energy, calculate the arrow's speed when it is released.

Answers on p. 141

Work and power

REVISED

Work

A force does work on an object when the force causes an object to move in the direction of the force.

work = force × distance moved in the direction of the force

$$W = Fs$$

When work is done by a force, the energy store of an object changes. For example:
- When 400 J of work is done to lift a box, the gravitational potential energy store of the box increases by 400 J.
- When 5000 J of work is done to accelerate a car, the kinetic energy store of the car increases by 5000 J.

work, W, in joules, J

force, F, in newtons, N

distance, s, in metres, m

Exam tip

Note, the equation for work will only be examined in the second paper. The equation is included here to help you with the definition of power.

Power

Power is the rate at which energy is transferred or the rate at which work is done.

$$\text{power} = \frac{\text{energy transferred}}{\text{time}}$$

or

$$\text{power} = \frac{\text{work done}}{\text{time}}$$

$$P = \frac{E}{t}$$

or

$$P = \frac{W}{t}$$

power, P, in watts, W

energy, E, transferred in joules, J

work done, W, in joules, J

time, t, in seconds, s

Now test yourself

11 State the correct unit for each of the following:
    (a) force
    (b) power
    (c) energy
    (d) work.
12 A weightlifter lifts a weightlifting bar of total mass 110 kg, from the ground above his head, to a height of 2.3 m. Gravitational field strength is 9.8 N/kg.
    (a) Calculate the increase in the gravitational potential energy store of the weightlifting bar.
    (b) The lift took 1.7 s. Calculate the average power developed to lift the weights.
13 An astronaut has a mass of 120 kg in his spacesuit. He needs to climb 8.0 m up a ladder into his spacecraft, which has landed on Mars. The gravitational field strength on Mars is 3.7 N/kg.
    (a) Calculate the astronaut's weight on Mars.
    (b) Calculate the increase to his gravitational potential energy store after climbing up the ladder.
    (c) Describe the energy transfers that take place as the astronaut climbs back into the spacecraft.
14 A crane lifts a weight of 15 000 N through a height of 28 m in 84 s. Calculate the output power of the crane in kW.
15 An electrical heater transfers 3000 J of thermal energy to a cup of water in 2 minutes. Calculate the power of the heater.

Answers on p. 142

Energy changes in systems

The amount of thermal energy stored in or released from a system as its temperature changes can be calculated using the equation:

change in thermal energy = mass × specific heat capacity × temperature change

$$\Delta E = mc\Delta\theta$$

change in thermal energy, ΔE, in joules, J

mass, m, in kilograms, kg

specific heat capacity, c, in joules per kilogram per degree Celsius, J/kg°C

temperature change, $\Delta\theta$, in degrees Celsius, °C

The specific heat capacity of a substance is the amount of energy required to raise the temperature of one kilogram of the substance by one degree Celsius.

Required practical 1

An investigation to measure the specific heat capacity of a material

There are several ways to measure specific heat capacity. All methods rely on the same principle: the decrease in one energy store (or work done) leads to the increase in thermal energy store of another material.

The apparatus shown in Figure 1.5 is used to warm up a steel block.

→

Exam tip

You should be able to describe an experiment and calculate specific heat capacity.

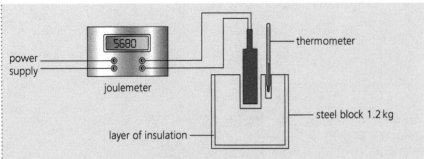

Figure 1.5

The following results were recorded:
- Electrical work done by the power supply to heat the steel was 5680 J.
- The initial temperature of the steel block was 18.2 °C.
- The final temperature of the steel block was 27.9 °C.
- Mass of the block was 1.2 kg.

$$\Delta E = mc\Delta\theta$$

$$\Delta\theta = 27.9 - 18.2 = 9.7\,°C$$

$$5680 = 1.2 \times c \times 9.7$$

$$c = \frac{5680}{1.2 \times 9.7}$$

$$= 490\,J/kg\,°C$$

The accepted value for the specific heat capacity of steel is 450 J/kg °C.

The measured value in a school laboratory is usually higher than the accepted value because:
- no account is made of thermal energy transfer to the surroundings
- the thermometer and heater require energy to warm them up too.

These are examples of systematic errors, because they always affect the answer by the same amount.

Now test yourself

16 State the unit for specific heat capacity.
17 Use the data below to answer the following questions.
    Specific heat capacity of water is 4200 J/kg °C
    Specific heat capacity of air is 1000 J/kg °C
    (a) Calculate the energy required to warm up 90 kg of air in a room from 7 °C to 23 °C.
    (b) A kettle is filled with 0.8 kg of water at a temperature of 18 °C. The kettle has a power rating of 2200 W and it is switched on for 2 minutes.
        (i)   Calculate the electrical work done by the kettle in two minutes.
        (ii) State the thermal energy transferred to the water; what assumptions are you making?
        (iii) Calculate the temperature of the water after 2 minutes of heating.
    (c) A heater transfers 12 000 J of energy to a block of copper of mass 1 kg. During the heating the temperature of the copper rises from 22 °C to 52 °C. Calculate the specific heat capacity of copper.
18 Your coffee has cooled down to a temperature of 30 °C, and you like to drink it at a temperature of 50 °C. Your coffee has a mass of 0.2 kg and a specific heat capacity of 4000 J/kg °C.
    (a) Your microwave oven has a power rating of 800 W. Calculate how long you need to heat the coffee to reach a temperature of 50 °C.
    (b) After heating, you find that the temperature is only 47 °C. Explain why the temperature is lower than your calculation predicted.

Answers on p. 142

Energy can be transferred usefully, stored or dissipated, but cannot be created or destroyed.

Storage and dissipation

You have already met the idea of energy being stored, for example in the kinetic energy of a moving car, or in the elastic potential energy of a stretched spring (page 6).

When energy is transferred from one store to another, energy can be dissipated or wasted. When energy is dissipated, it is stored in less useful ways.

> **Example**
>
> Chemical energy is stored in petrol; we want that energy to do work against resistive forces to keep a car moving. But when the petrol burns, energy is also transferred to the thermal energy store of the surroundings. That energy is wasted, as we did not want to produce it, and we cannot recapture it to do anything useful.

Energy saving

Wherever possible, we try to avoid the **dissipation** of energy, so that we maximise the useful transfer of energy from one store to another.

> **Dissipation** is the wasting and spreading out of thermal energy into the surroundings.

In the home

When we heat our home, eventually all the energy in the home's thermal store will be dissipated to the outside of the house. We get the best value for our heating bill, and we avoid wasting energy resources, if we slow down the process of thermal energy transfer.
- We insulate the loft.
- Carpets insulate the floors.
- Windows and doors are draught proofed.
- Thick walls reduce energy transfer.
- The cavity between the inside and outside wall of the house can be insulated.
- Double glazing reduces energy transfer through the windows, by trapping a layer of air between two panes of glass.

Insulation usually involved trapping a layer of air in fibres. Air carries energy efficiently by convection, but when air is trapped it is a good insulator, because air has a very low **thermal conductivity**.

> The **thermal conductivity** of a material is a measure of how quickly energy is transferred by conduction through it. Metals have very high conductivities. Brick and glass have lower conductivities, but energy still flows through them fast enough to cool a house down on a cold day.

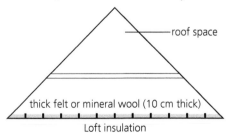

Figure 1.6 **Loft insulation.**

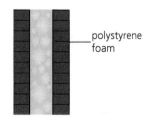

Figure 1.7 **Polystyrene foam acts as an insulator in the cavity between the walls.**

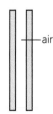

Figure 1.8 **Air acts as an insulator in double-glazed windows.**

Car design

We try to get the most useful energy out of petrol that we can.

- Cars are fuel efficient, which means less energy is dissipated to the surroundings.
- Cars are streamlined to reduce air resistance.
- Moving parts are lubricated with oil to reduce friction.

By reducing the dissipation of energy we increase the efficiency of the intended energy transfer. In the example of the car, streamlining and lubrication increase the amount of energy available to do work against resistive forces.

Required practical 2

Investigating thermal insulation

Thermal insulation helps to keep things warm (or cool) for longer by reducing unwanted transfers of energy in a system.

You need to be able to design and discuss fair tests to evaluate the effectiveness of various insulators, for example, to prevent heat loss from a beaker of hot water.

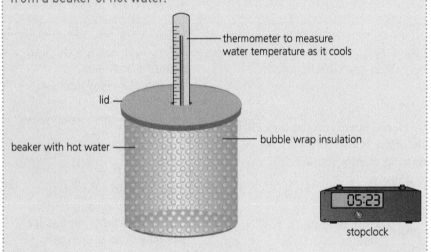

Figure 1.9

Method

1. A measured quantity of water (e.g. 200 ml) at a temperature of 70 °C is poured into a beaker.
2. The temperature of the water can be measured after 10 minutes to assess the energy transfer to the surroundings.
3. Using the same volume of water, at the same initial temperature, the final temperature can be measured again after 10 minutes for the following changes:
   - a lid is placed on the beaker
   - various insulators are used to wrap the sides of the beaker – for example, bubble wrap, cloth, corrugated cardboard.

Efficiency

The energy efficiency for any energy transfer can be calculated using the equation:

$$\text{efficiency} = \frac{\text{useful output energy transfer}}{\text{total input energy transfer}}$$

Efficiency can also be calculated using the equation:

$$\text{efficiency} = \frac{\text{useful power output}}{\text{total power input}}$$

Example

A steam engine uses coal as its fuel. When the chemical store of the coal transfers 200 kJ of energy 24 kJ of work is done against resistive forces. Calculate the efficiency.

Answer

$$\text{efficiency} = \frac{\text{useful output energy transfer}}{\text{total input energy transfer}}$$

$$= \frac{24}{200}$$

$$= 0.12 \text{ or } 12\%$$

Exam tip

You can express efficiency as a decimal or as a percentage. Since efficiency is a ratio of energies (or powers) it has no unit.

Now test yourself

TESTED

19 Explain the meaning of the phrase *energy dissipation*.
20 (a) List three ways in which we reduce energy losses from our homes.
   (b) Explain how each of your choices reduces energy dissipation.
21 Explain how a car can be designed to be more efficient in its use of petrol.
22 This question refers to Required practical 2 on page 12.
   (a) Explain how you would ensure the investigation used fair tests.
   (b) Explain how you would use your results to show which is the best insulator for the sides of the beaker.
   (c) Identify a possible source of a random error in this experiment.
23 Define *efficiency*.
24 A car is supplied with 10 kg of fuel. A kilogram of fuel stores 4.5 MJ of chemical energy. The efficiency of the car is 30%.
   (a) Calculate the amount of energy available for useful work against resistive forces.
   (b) Calculate the amount of energy dissipated after all the fuel has been used.
25 The human body is about 25% efficient in transferring energy from the body's chemical store, to allow the body to do mechanical work.
   (a) A boy of mass 60 kg climbs a tower of height 35 m. Calculate the energy in his gravitational potential store after the climb is completed. Gravitational field strength is 9.8 N/kg.
   (b) Calculate the amount of chemical energy transferred for the boy's climb.
   (c) Calculate the energy dissipated during the boy's climb.
   (d) Assuming all the dissipated energy is transferred to the boy's thermal store, calculate the increase in his temperature. The specific heat capacity of a boy is 4000 J/kg °C.

Answers on pp. 142–143

National and global energy resources

When we discuss energy resources, we are often interested in stores of energy that we can use to generate electrical power.

An energy store sets in motion a resource such as a gas or water that moves past a turbine which drives an electrical generator.

Non-renewable energy resources

Much of our electricity in the UK comes from the fossil fuels, coal, oil and gas. These fuels store chemical energy. To release the energy, the fossil fuels must be burned. Once the fuels are burned, they are gone forever, because these fuels have taken millions of years to form. Nuclear fuels, uranium and plutonium, are also non-renewable, but there is a plentiful supply. These fuels are described as **non-renewable energy resources**, because there is a finite supply.

> **Non-renewable energy resources** will run out, because there are finite reserves, which cannot be replenished.
>
> **Renewable energy resources** will never run out, because these can be replenished.

Renewable energy resources

By contrast **renewable energy resources** will never run out. We obtain renewable energy from the Sun, tides, waves, rivers and waterfalls, from the wind and from thermal energy in the Earth itself.

Using fuels

The main uses of fuels are as follows:
- **Electricity**. In the UK electricity is generated using different energy resources: just over half is generated from fossil fuels, about 20% is generated from nuclear fuel and the rest is made using renewable resources.
- **Transport**. Fuels such as petrol, diesel and kerosene are produced from oil. These fuels drive our cars, trains and planes. Electricity is also used to run trains, and rechargeable batteries in electric cars are used.
- **Heating**. Most of our home heating is provided by gas and electricity. Some homes have oil-fired boilers, or burn solid fuels such as coal and wood.

> **Exam tip**
>
> Common nuclear fuels are uranium and plutonium.

Reliability of energy resources

To generate electricity we need reliable energy resources. Fossil fuels are reliable, as we can mine coal and extract oil and gas from wells. However, our fuel resources might run out in the next hundred years, or become very expensive.

Tidal power is reliable, because we have high tides twice a day. However, the times of the high tides change each day, so the peak of electricity generation might not coincide with peak demand.

Solar, wind and hydroelectric power make useful contributions to electricity generation in many countries. But these are not reliable: there is less solar energy available in winter or on cloudy days; wind strength varies considerably; in some countries there is less hydroelectric power available in winter as rivers freeze.

> **Exam tip**
>
> Make sure you understand the advantages and disadvantages of resources used to generate electricity.

Environmental issues

There is an increasing amount of evidence to show that the Earth is warming up (global warming). It seems likely that the Earth will soon be about 2 °C warmer (on average) than it was 50 years ago. Many people think that global warming is linked to the production of carbon dioxide and other **greenhouse gases** that trap radiation in the Earth's atmosphere.

As a result of global warming, many countries are committed to reducing the use of fossil fuels, and want to generate their electricity using renewable energy resources.

Renewable energy resources also have their impact on the environment.
- Wind turbines can be noisy and people object to them spoiling the look of the countryside.
- Tidal barrages that trap water at high tides can affect the habitat of wildlife.
- Hydroelectric dams affect the flow of rivers, and lakes made behind dams have flooded small towns, causing communities to be relocated.

The production of electricity requires a balance between our needs and environmental issues.

> A **greenhouse gas** is one that traps radiation in the Earth's atmosphere and, therefore, contributes to global warming. Carbon dioxide and methane are examples of greenhouse gases.

Political and economic issues

We can only solve the problems caused by the production of greenhouse gases if all countries agree. The UK plans to stop producing electricity from coal by 2025. However, unless all countries do the same, there will still be problems with global warming.

One reason countries continue to burn coal is cost. Coal-fired power stations are cheaper to run than nuclear ones, or renewable energy resources. To solve environmental issues, we need to be prepared to pay more for our electricity.

Now test yourself

TESTED

26 (a) Explain what a renewable energy resource is. Give an example.
  (b) Explain what a non-renewable energy resource is. Give an example.
27 (a) State two advantages of using a coal-fired power station.
  (b) State two environmental problems associated with coal-fired power stations.

Answers on p. 143

Summary

- Energy is an idea that cannot be described by a single process. However, we pay an enormous amount of attention to energy because it is conserved.
- There are different stores of energy. In any process, energy can be transferred from one store to another, but energy is never created or destroyed.
- Energy stores include: kinetic, chemical, internal (or thermal), gravitational potential, magnetic, electrostatic, elastic potential and nuclear.
- Energy can be transferred from one store to another by: mechanical work, electrical work, heating and radiation (mechanical and electromagnetic).

→

- The amount of energy transferred to or from a store may be calculated from these equations:

  kinetic energy $E_k = \frac{1}{2}mv^2$

  gravitational potential energy: $E_p = mgh$

  elastic potential energy: $E_e = \frac{1}{2}ke^2$

  thermal energy: $\Delta E = mc\Delta\theta$
- Mechanical work done is calculated using:
  $W = Fs$
- Power is calculated using the equations:

  $P = \dfrac{\text{energy}}{\text{time}}$

  or

  $P = \dfrac{\text{work}}{\text{time}}$
- Units:
  - power – watts (W)
  - energy – joules (J)
  - work – joules (J)
  - force – newtons (N)

- In many processes there are unwanted energy losses in the form of the transfer of thermal energy. This energy cannot be recovered into a useful form.

  $\text{efficiency} = \dfrac{\text{useful output energy transfer}}{\text{useful input energy transfer}}$

  or

  $\text{efficiency} = \dfrac{\text{useful power output}}{\text{total power input}}$
- Energy resources can be non-renewable – oil, gas, coal and nuclear fuels.
  Or, energy resources can be renewable – wind, waves, hydroelectric, geothermal, solar, for example.
- Fuels are used for transport, heating and generating electricity.

Exam practice

1 Which of the following is required for a hydroelectric power station?
   A Sunlight
   B A supply of falling water
   C A supply of hot water from the Earth   [1]
2 An electric car uses a battery to power it. The car accelerates from rest.
   (a) Choose words from the list below that describe the energy transfers while the car is accelerating. [3]

   kinetic energy   gravitational   potential   thermal   chemical   elastic potential

   The battery has a store of .... energy. When the car accelerates, the motor transfers some
   useful energy to increase the car's store of .... energy. Some energy is wasted which increases
   the .... energy store of the motor and is surroundings.

   (b) When the car is accelerating the motor's output power is 4800 W. The motor transfers energy
       into useful energy at a rate of 1200 W.
       Calculate the efficiency of the car's motor.   [2]
3 (a) State one advantage and one disadvantage of using nuclear power.   [2]
   (b) A nuclear power station has a power output of 3000 MW. A wind turbine has a maximum power
       output of 2 MW.
       (i)   How many watts, W, are there in a megawatt, MW?   [1]
       (ii) How many wind turbines, working at their maximum rate, produce the same power as a
             nuclear power station?   [2]
       (iii) Explain one advantage and one disadvantage of using wind turbines instead of a nuclear
             power station.   [2]
4 (a) (i) Name one renewable fuel and one non-renewable fuel.   [2]
       (ii) Explain what the words renewable and non-renewable mean in this context. [2]
   (b) Explain two reasons why governments may wish to increase the amount of electricity
       generated using renewable energy resources.   [2]
5 When we heat our homes, often energy is wasted.
   Choose three ways in which energy can be wasted, and explain how that waste can be reduced. [6]

6 A heater is used to increase the temperature of a block of tin. The graph in Figure 1.11 shows how the temperature of the tin rises, with the energy transferred by heating. The energy is measured by the joulemeter.

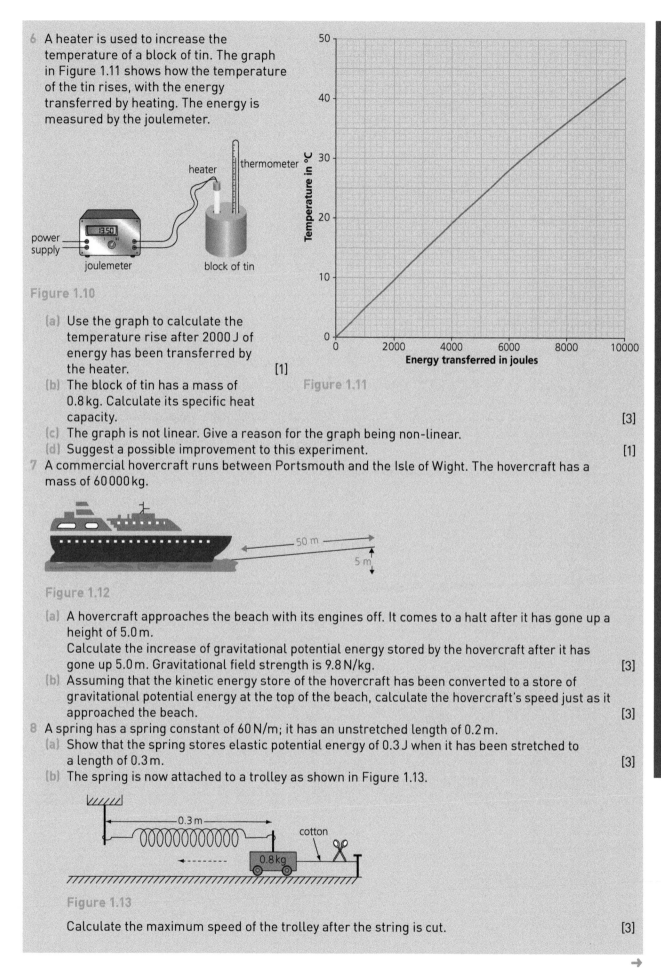

Figure 1.10

Figure 1.11

(a) Use the graph to calculate the temperature rise after 2000 J of energy has been transferred by the heater. [1]

(b) The block of tin has a mass of 0.8 kg. Calculate its specific heat capacity. [3]

(c) The graph is not linear. Give a reason for the graph being non-linear.

(d) Suggest a possible improvement to this experiment. [1]

7 A commercial hovercraft runs between Portsmouth and the Isle of Wight. The hovercraft has a mass of 60 000 kg.

Figure 1.12

(a) A hovercraft approaches the beach with its engines off. It comes to a halt after it has gone up a height of 5.0 m.
Calculate the increase of gravitational potential energy stored by the hovercraft after it has gone up 5.0 m. Gravitational field strength is 9.8 N/kg. [3]

(b) Assuming that the kinetic energy store of the hovercraft has been converted to a store of gravitational potential energy at the top of the beach, calculate the hovercraft's speed just as it approached the beach. [3]

8 A spring has a spring constant of 60 N/m; it has an unstretched length of 0.2 m.

(a) Show that the spring stores elastic potential energy of 0.3 J when it has been stretched to a length of 0.3 m. [3]

(b) The spring is now attached to a trolley as shown in Figure 1.13.

Figure 1.13

Calculate the maximum speed of the trolley after the string is cut. [3]

9 A bullet has a mass of 0.02 kg and travels with a speed of 400 m/s.

(a) Calculate the kinetic energy stored in the bullet. [3]

(b) The bullet hits a tree and travels a depth of 0.25 m into the tree. A resistive force does work to slow down the bullet. Calculate the size of this force. [3]

(c) The kinetic energy of the bullet is transferred to thermal energy and the bullet's temperature rises. The bullet has a specific heat capacity of 500 J/kg °C. Calculate its temperature rise. [3]

10 An electric winch is used to pull up a truck, as shown in Figure 1.14.

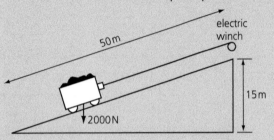

Figure 1.14

(a) Calculate the gain in gravitational potential energy of the truck after it has been pulled up 15 m. [3]

(b) The winch uses a 5 kW electric supply and takes 12 s to pull the truck 50 m along the slope. Calculate the electrical work done by the winch. [3]

(c) Calculate the efficiency of the winch. [2]

11 A power supply is used to heat 0.1 kg of water in an insulated beaker. The water has a temperature of 20 °C.

Use the information in the diagram to calculate how long it takes for the water to warm to 50 °C. Water has a specific heat capacity of 4200 J/kg °C. [6]

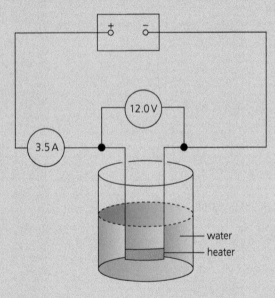

Figure 1.15

Answers and quick quiz 1 online

ONLINE

2 Electricity

Electrical power is an integral part of our lives. It fills our world with artificial light and information, and allows us to be entertained at any time of the day.

Current, potential difference and resistance

Figure 2.1 shows the standard **circuit symbols** you need to know.

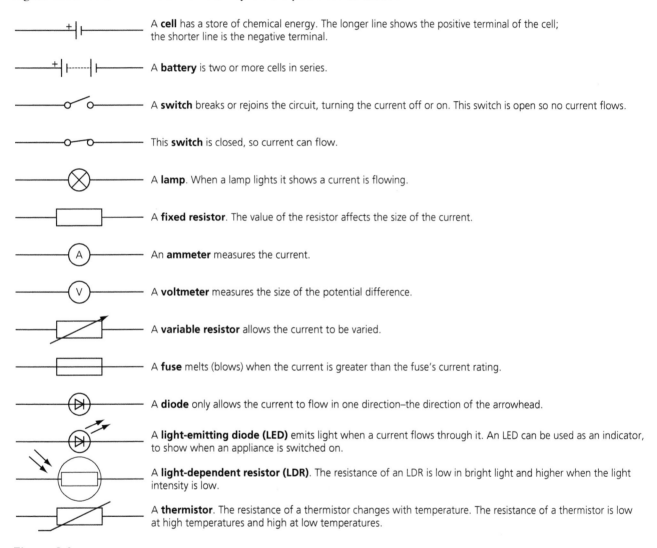

A **cell** has a store of chemical energy. The longer line shows the positive terminal of the cell; the shorter line is the negative terminal.

A **battery** is two or more cells in series.

A **switch** breaks or rejoins the circuit, turning the current off or on. This switch is open so no current flows.

This **switch** is closed, so current can flow.

A **lamp**. When a lamp lights it shows a current is flowing.

A **fixed resistor**. The value of the resistor affects the size of the current.

An **ammeter** measures the current.

A **voltmeter** measures the size of the potential difference.

A **variable resistor** allows the current to be varied.

A **fuse** melts (blows) when the current is greater than the fuse's current rating.

A **diode** only allows the current to flow in one direction–the direction of the arrowhead.

A **light-emitting diode (LED)** emits light when a current flows through it. An LED can be used as an indicator, to show when an appliance is switched on.

A **light-dependent resistor (LDR)**. The resistance of an LDR is low in bright light and higher when the light intensity is low.

A **thermistor**. The resistance of a thermistor changes with temperature. The resistance of a thermistor is low at high temperatures and high at low temperatures.

Figure 2.1

Electrical charge and current

Figure 2.2 shows a simple circuit.

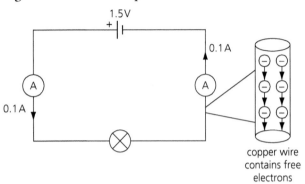

Figure 2.2

In this circuit a cell provides a potential difference (p.d.) of 1.5 V to drive a current of 0.1 A.

- The potential difference is a measure of the electrical work done by the cell to drive the current round the circuit.
- The current is a measure of the rate at which charge flows round the circuit.
- The charge is measured in coulombs.
- We have a convention that current flows from positive to the negative terminal of the cell. But when electrons flow, they travel from the negative to the positive terminal of the cell.

Charge, current and time are linked by this equation:

charge flow = current × time

$$Q = It$$

charge flow, Q, in coulombs, C

current, I, in amperes, A (amp is acceptable for ampere)

time, t, in seconds, s

Small currents are measured in milliamps (mA)

Now test yourself

TESTED

1 Draw a circuit diagram to show a cell, an ammeter, a lamp and a resistance connected in series.
2 In an electrical circuit, a charge of 12 C flows round a circuit in 2 minutes. Calculate the current in
   (a) amps
   (b) milliamps.

Answers on p. 143

Current, resistance and potential difference

The current, I, through a component depends both on the resistance, R, of the component and the potential difference, V, across the component. The greater the resistance of the component, the smaller the current, for a particular potential difference.

The current, potential difference and resistance are linked by the equation:

potential difference = current × resistance

$$V = IR$$

Large resistances may be measured in kilohms (kΩ) and megohms (MΩ).

- $1\,k\Omega = 1000\,\Omega$
- $1\,M\Omega = 1\,000\,000\,\Omega$

potential difference, V, in volts

current, I, in amperes (or amps)

resistance, R, in ohms, Ω

Ammeters and voltmeters

Figure 2.3 shows how an ammeter and voltmeter are connected to measure resistance. The ammeter must be in series with the resistor and the voltmeter must be in parallel with the resistor.

Example

In Figure 2.3 the voltmeter reads 12 V and the ammeter reads 0.06 A. Calculate the resistance.

Answer

$$V = IR$$
$$12 = 0.06 \times R$$
$$R = \frac{12}{0.06}$$
$$= 200\,\Omega$$

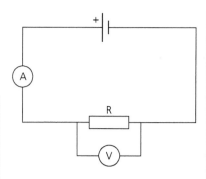

Figure 2.3

Required practical 3

You should have set up a circuit to investigate how the resistance of a given wire depends on its length. Figure 2.4 shows you the circuit to use.

You will have found that **the resistance of the wire is proportional to its length**.

For example: if a wire of length 40 cm has a resistance of 7.5 Ω, a length of 80 cm of the same wire has a resistance of 15.0 Ω.

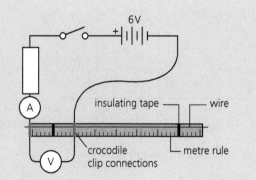

Figure 2.4

You are also expected to be able to set up circuits to investigate combinations of resistors in series and in parallel. This is covered in Exam practice question 2 on pages 31–32.

Resistors

Ohmic conductors and non-ohmic conductors

The current through an ohmic conductor (at a constant temperature) is directly proportional to the potential difference across it. A graph of current against potential difference is a straight line (Figure 2.5). The resistance stays the same as the current changes.

The resistances of components such as **lamps, diodes, thermistors** and **light-dependent resistors** are not constant; the resistance changes with the current through the component. These are **non-ohmic conductors**.

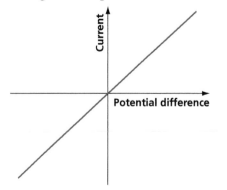

Figure 2.5 The *I–V* graph for an ohmic conductor at a constant temperature.

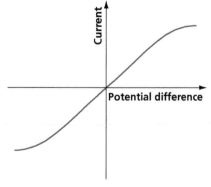

Figure 2.6 The current in a filament lamp is not proportional to the potential difference.

A filament lamp

When a current flows through a filament lamp, the filament heats up.
The resistance of a filament lamp increases as the temperature increases
(Figure 2.6).

A diode

A diode is a component that allows current to flow only one way (Figure 2.7).
A diode has a very high resistance in the reverse direction.

Light-dependent resistor (LDR)

The resistance of an LDR decreases as the light intensity increases (Figure 2.8).

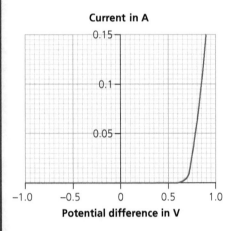

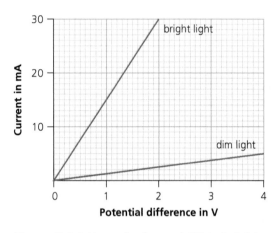

Figure 2.7 An *I–V* graph for a diode.

Figure 2.8 *I–V* graphs for an LDR in bright and dim light.

Thermistor

The resistance of a thermistor decreases as the temperature rises (Figure 2.9).

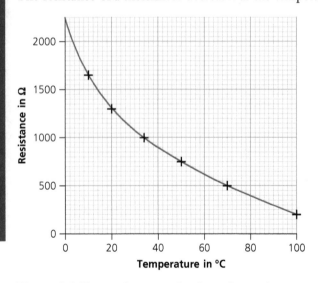

Figure 2.9 The resistance of a thermistor changes with the temperature.

Required practical 4

You should be able to describe an experimental set-up and procedure that enables you to investigate the *I–V* characteristic graphs for a filament lamp, a diode and a resistance at a constant temperature.

This can be done using the apparatus shown in Figure 2.10.
- The current and potential difference values are recorded by reading the ammeter and voltmeter.
- The current can be altered by changing the number of cells or by changing the variable resistor.
- Using your data, you then plot an *I–V* graph for each component.

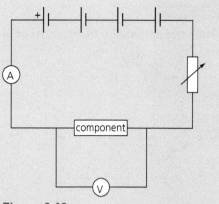

Figure 2.10

Now test yourself

TESTED

3 (a) Explain what is meant by an *ohmic resistor*.
  (b) With reference to Figure 2.6, explain why a filament lamp is a non-ohmic resistor.
4 This question refers to a diode with the *I–V* characteristics shown in Figure 2.7.
  (a) Use the graph to calculate the potential difference across the diode, when a current of 0.1 A flows through it.
  (b) Calculate the resistance of the diode when 0.1 A is flowing through it.
5 In the list below there are five units for different electrical quantities.
  (a) Which is the correct unit for resistance?
  (b) Which is the correct unit for electrical charge?

  volt coulomb amp ohm watt

6 Describe an experiment to determine the *I–V* characteristics for a filament lamp. In your explanation you should state what apparatus you will use, and how you will use it. You should also explain what measurements you will take.

Answers on p. 143

Series and parallel circuits

There are two ways of connecting electrical components in a circuit, in **series** and in **parallel**.

Series circuits

For components connected in series:
- there is the same current through each component
- the potential difference of the power supply is shared between the components. If there are just two components then:

$V_{supply} = V_1 + V_2$

- the total resistance of two components is the sum of the resistance of each component:

$R_{total} = R_1 + R_2$

Example

Calculate the total resistance between A and B.

Figure 2.11

Answer

$= 5 + 10 = 15\,\Omega$

Example

State the potential difference across lamp 2 in Figure 2.12.

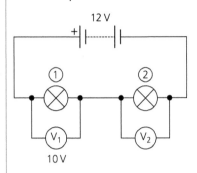

Figure 2.12

Answer

$$V_{supply} = V_1 + V_2$$

$$12 = 10 + V_2$$

$$V_2 = 12 - 10 = 2\,V$$

Parallel circuits

For components connected in parallel:
- the potential difference across each component is the same
- the total current through the whole circuit is the sum of the currents through the separate components
- the total resistance of two resistors in parallel is less than the resistance of the smaller individual resistor.

Example

In Figure 2.13, state the value of the current going through lamp 2, and the potential difference across each lamp.

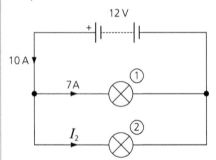

Figure 2.13

Answer

Current:

$$I_2 = 10 - 7 = 3\,A$$

Each lamp has 12 V across it.

Now test yourself

TESTED

7 Calculate the resistance between:
  (a) AB
  (b) CD.

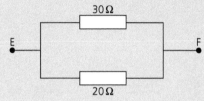

Figure 2.14

8 Which of the following correctly states the resistance between points E and F?

Figure 2.15

50 Ω   25 Ω   less than 20 Ω   between 30 Ω and 50 Ω

9 (a) Calculate the potential difference V_2 in Figure 2.16.
  (b) Calculate the two ammeter readings A_1 and A_2 in Figure 2.17.

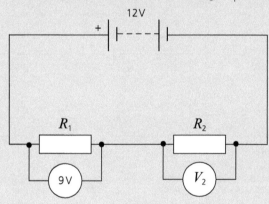

Figure 2.16

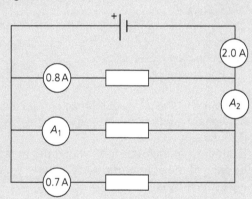

Figure 2.17

Answers on p. 143

Circuit calculations

You will be expected to use the rules about series and parallel circuits (above) to solve circuit problems. Some examples are given below.

Example

Calculate the readings on:
1 the ammeter, A
2 the voltmeter, V.

Answer
1 $V = IR$
   $4 = I \times 6$
   $I = \dfrac{4}{6}$
     $= 0.67\,A$

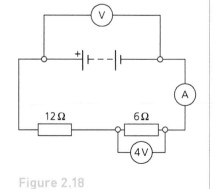

Figure 2.18

Typical mistake

Sometimes students combine the wrong p.d. with the wrong resistor. Remember, when you use $I = \dfrac{V}{R}$ to calculate the current, V is the p.d. across the resistor R: in Figure 2.18, 4V across the 6Ω.

2 The p.d. across the 12 Ω resistor is:

$V = IR$

$= 0.67 \times 12$

$= 8\,V$

So the battery p.d., as measured by the voltmeter V, is: 8 + 4 = 12 V

Now test yourself

10 (a) A thermistor is connected into the circuit shown in Figure 2.19, when the temperature is 15 °C.

Use the information in the diagram to calculate
(i) the potential difference measured by the voltmeter, V
(ii) the resistance of the thermistor.

(b) The next day the temperature goes up to 25 °C. Explain what happens to the voltmeter reading.

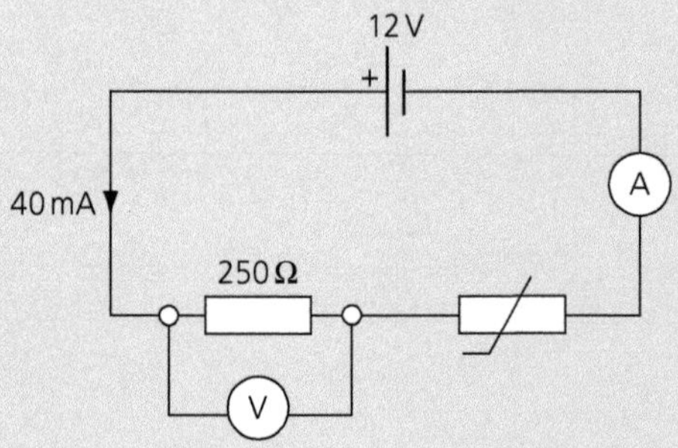

11 In Figure 2.20, the switch S is left open.

(a) State the currents measured by the ammeters, A_1 and A_2.

(b) Use the information in the diagram to calculate the resistance, R.

(c) The switch is now closed, and the ammeter A_1 reads 0.1 A.
State the new readings on:
(i) the ammeter, A_2
(ii) the voltmeter across the battery.

(d) Explain why the total resistance of the circuit between the points AB is less when the switch is closed.

(e) Explain what happens to the currents measured by each of the ammeters, A_1, A_2 and A_3, when the light intensity is increased.

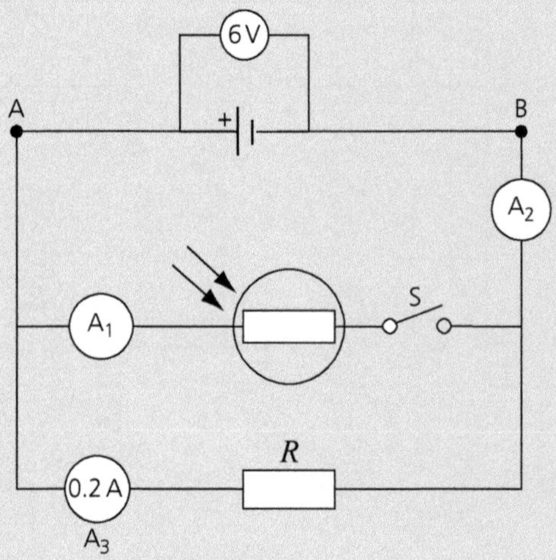

Figure 2.19

Figure 2.20

Answers on p. 143

Domestic use and safety

Direct and alternating potential difference

Power supplies can provide direct or alternating potential differences. This is illustrated in Figure 2.21.

- The blue line shows a direct potential difference of 6 V. This will make a direct current (d.c.) flow in one direction through a resistor.
- The red line shows an alternating potential difference of 6 V. This changes direction so an **alternating current** (a.c.) flows first one way, then the other through a resistor. The peak value of the 6 V a.c. supply rises above 6 V to make up for the time when the potential difference is close to zero.

> **Alternating current (a.c.)** is current that flows one way and then the other.

Mains supply

The mains supply in the United Kingdom has a frequency of 50 Hz and a potential difference of about 230 V. A frequency of 50 Hz means that one cycle – as shown by the red curve in Figure 2.21, takes one-fiftieth of a second.

Mains electricity

Most electrical appliances are connected to the mains using a three-core cable. Each of the wires inside the cable is colour coded for easy identification.
- Live wire – brown
- Neutral wire – blue
- Earth wire – green and yellow stripes

The live wire carries the alternating potential difference from the mains supply. The neutral wire completes the circuit. So the live and neutral wires carry the current to and from an electrical appliance.

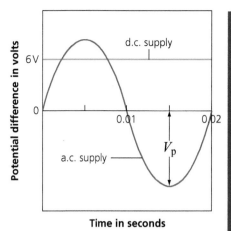

Figure 2.21 V_p is the peak voltage.

Figure 2.22 A cable has three wires; earth (green/yellow), neutral (blue), live (brown).

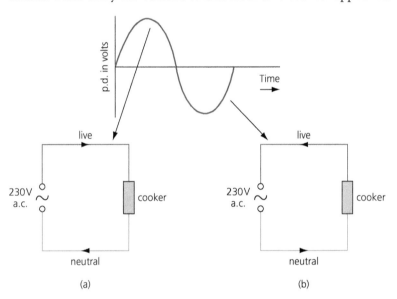

Figure 2.23

- The potential difference between the live wire and earth (0 V) is about 230 V. Even though an appliance is off and there is no current in the mains circuit, a live wire is dangerous. If you touch a live wire, current passes through you to earth, giving you a painful shock.
- The neutral wire is close to earth potential (0 V).
- The earth wire is at 0 V, and only carries a current if there is a fault.

Earthing

Any electrical appliance that has a metal case should be earthed. The toaster in Figure 2.24 has the earth wire connected to its metal case.

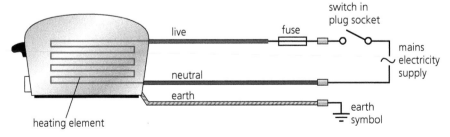

Figure 2.24

Any contact between the live wire and earth is potentially dangerous, because a large current passes to earth, which could start a fire.

Energy transfers

Power

The electrical power transferred by any electrical device is equal to the energy transferred per second.

The power transferred depends on the potential difference and current:

$$P = VI$$

power = (current)² × resistance

$$P = I^2R$$

> power, P, in watts, W
>
> potential difference, V, in volts, V
>
> current, I, in amperes, A (or amps)
>
> resistance, R, in ohms, Ω

Energy

When charge flows round a circuit, electrical work is done.

The energy transferred by electrical work depends on how long the appliance is switched on.

energy transferred = power × time

$$E = Pt$$

energy transferred = charge flow × potential difference

$$E = QV$$

> energy transferred, E, in joules, J
>
> time, t, in seconds, s
>
> charge flow, Q, in coulombs, C
>
> potential difference, V, in volts V

Examples

1 The information plate on a kettle is marked as follows:

    230 V   50 Hz   2650 W

  Calculate the current drawn from the supply.
2 Calculate the energy transferred by the kettle (in Example 1) in 2 minutes.
3 When 150 C of charge passes through a battery, 900 J of energy is transferred. Calculate the potential difference of the battery.

Answers

1   $P = VI$

  $2650 = 230I$

  $I = \dfrac{2650}{230}$

    $= 11.5\,\text{A}$

2 $E = Pt$

    $= 2650 \times 120$

    $= 318\,000\,\text{J}$

    $= 318\,\text{kJ} \approx 320\,\text{kJ}$

3   $E = QV$

  $900 = 150 \times V$

  $V = 6\,\text{V}$

Typical mistake

Often power and energy are confused.

$$\text{power} = \frac{\text{energy}}{\text{time}}$$

Power is measured in joules per second or watts.

Energy is measured in joules.

Typical mistake

When using time in an equation, remember to turn the time into **seconds**.

Now test yourself

12 Which of the following is the correct unit for:
   (a) power
   (b) energy
   (c) charge?
   volt   amp   joule   ohm   watt   coulomb

13 Which of the following is an equivalent unit for:
   (a) volt
   (b) amp?
   joule × coulomb   coulomb/second   joule/coulomb   joule/second   coulomb × second

14 Calculate the power rating in watts of:
   (a) A fire that draws 8 A from a 230 V supply.
   (b) A lamp that draws 5 A from a 12 V supply.

15 An electric shower runs from a 230 V 15 A supply. Calculate the energy transferred to heat the water when someone has a shower for 3 minutes.

16 Calculate the electrical work done when:
   (a) 150 C of charge flows through a lamp with a potential difference of 12 V across it
   (b) a current of 8 mA flows through a 6 V battery charger for 4 hours.

17 A current of 0.1 A flows through a resistor of 220 Ω for 20 minutes. Calculate the energy transferred by the resistor.

Answers on pp. 143–144

The National Grid

The National Grid is a system of cables and transformers that links power stations to consumers.

Electrical power is transferred through the National Grid.

Transformers step up the potential difference from the power station to the transmission cables (seen on overhead power lines). The high potential difference makes the current much lower; therefore less energy is wasted when the current is carried over long distances. Transformers in towns step down the high potential difference to the safe 230 V we use in our homes. This topic is covered in greater detail in Chapter 7, page 127.

Static electricity

When electric charge flows through a wire, there is a current. However, there are many occasions when electric charge is stationary (or static) on the surface of an object. This charge is called static electricity.

Making static

Static electricity can be produced by rubbing some insulating materials together. Negatively charged electrons are rubbed off one material and onto another. The material that gains electrons becomes negatively charged. The material that loses electrons is left with an equal positive charge.

Sparks

When static electricity is produced, there is a potential difference between positive and negative charges. Sometimes this potential difference is so big that a spark crosses between the two types of charge. Then a current flows.

> **Exam tip**
>
> Always write in terms of electron flow when answering questions about static charge.

- You might have noticed small sparks when you pull off an item of clothing.
- Lightning is a dramatic example of static electricity. The bottom of a thundercloud becomes negatively charged. The potential difference between the cloud and Earth is many millions of volts. The cloud is discharged by a lighting flash (large spark).

Forces on charges

Simple experiments show that charged objects exert noticeable forces on each other when brought close together. The electrostatic force is a non-contact force.

- Two objects that carry the same type of charge repel each other.
- Two objects that carry opposite types of charge attract each other.

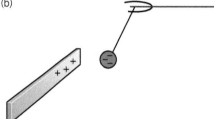

Figure 2.25

Electric fields

- A charged object creates an electric field around itself.
- The electric field is strongest close to the charged object.
- The direction of the field is the direction of the force on a positively charged object.

Figure 2.26 shows the electric field close to positively and negatively charged spheres.

- The field lines point away from the positively charged sphere.
- The field lines point towards the negatively charged sphere.
- As the distance increases away from the spheres, the field lines get further apart. This shows the field gets weaker away from the spheres.

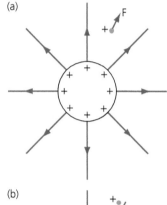
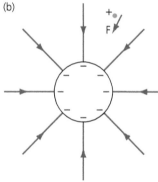

Figure 2.26

Now test yourself

TESTED

18 (a) A plastic rod is rubbed by a duster and the rod becomes negatively charged. Explain what has happened to the rod.
　　(b) What sign of charge is there on the duster?
19　Explain what is meant by the term *electric field*. Illustrate your answer with a diagram.

Answers on p. 144

Summary

- You should know the circuit symbols shown on page 19.
- Current is a flow of charge.

  charge = current × time

  $$Q = It$$

- Current is measured in amps, A.
- Charge is measured in coulombs, C.
- Potential difference is measured in volts, V.

  potential difference = current × resistance

  $$V = IR$$

- Resistance is measured in ohms, Ω.
- When two components are in series:
  - the current is the same through each component
  - the potential difference of the power supply is shared between each component
  - the total resistance is the sum of the two resistances,

  $$R_{total} = R_1 + R_2$$

- When two components are connected in parallel:
  - the potential difference across each component is the same
  - the total current through the circuit is the sum of the currents through each component,

  $$I_{total} = I_1 + I_2$$

  - the total resistance is less than the resistance of the smallest individual resistor.
- A direct current flows one way round a circuit. An alternating current switches from one direction to the other.
- power = potential difference × current

  $$P = VI$$

  $$P = I^2R$$

- energy transferred = power × time

  $$E = Pt$$

- energy transferred = charge flow × potential difference

  $$E = QV$$

- Static electricity is produced by transferring electrons from one insulator to another. One material loses electrons and becomes positively charged; the material gaining electrons acquires an equal negative charge.

Exam practice

1 Figure 2.27 shows an electrical circuit.

(a) The two cells are identical. State the potential difference of one cell. [1]

(b) State the reading on the voltmeter. [1]

(c) State the reading on the ammeter. [1]

(d) Show by calculation that the resistance of the lamp is about 12 Ω. [3]

(e) Which of the following best describes the total resistance of the circuit? [1]

27 Ω   between 15 Ω and 12 Ω   less than 12 Ω

2 A student designs an experiment to investigate the effect of adding two resistors together in series. His circuits are shown in Figure 2.28.

(a) Use the information in Figures 2.28 (a) and 2.28 (b) to show that the values of the resistors R_A and R_B are:

(i) $R_A = 20\,Ω$

(ii) $R_B = 40\,Ω$. [3]

(b) The resistors are now put in series as shown in Figure 2.28 (c). The student predicts that the current in this circuit will be 0.10 A. Show by calculation how the student reached this hypothesis. [2]

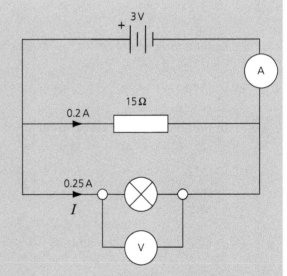

Figure 2.27

(c) The student puts the two resistors in parallel as shown in Figure 2.28 (d).
He now predicts that the total resistance of the circuit will be less than 20 Ω.
Use the information in Figures 2.28 (a) and (b) to explain how the student reached
his hypothesis. [3]

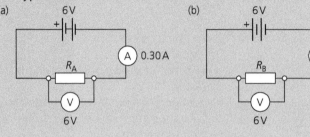

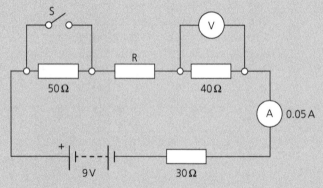

Figure 2.28

3 In Figure 2.29 the 9 V battery supplies a direct current of 0.05 A to the circuit.

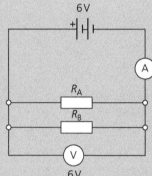

Figure 2.29

(a) Explain what is meant by *direct p.d.* [1]
(b) Use the information in the diagram to calculate the total resistance of the circuit. [3]
(c) Use your answer to part (b) to calculate the resistance, R. [1]
(d) The switch S is now closed. Explain what happens to each of the following (a calculation is not needed):
  (i) the total resistance of the circuit [1]
  (ii) the reading on the ammeter [1]
  (iii) the reading on the voltmeter. [1]

4 (a) Draw a circuit that you would use to obtain the data needed to draw a current–potential difference graph for a lamp. [3]
 (b) Sketch I–V graphs for:
  (i) a filament lamp [1]
  (ii) a resistor at a constant temperature [1]
  (iii) a diode. [1]

5 Figure 2.30 shows three resistors connected across a 12 V cell.

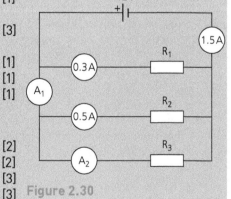

(a) Calculate the currents through the ammeters, A_1 and A_2. [2]
(b) Which resistance is greater, R_1 or R_2? Explain your answer. [2]
(c) (i) Calculate the resistance, R_2. [3]
  (ii) Calculate the power dissipated in the resistance, R_2. [3]

Figure 2.30

6 Figure 2.31 shows a circuit which includes a fixed resistor of 240 Ω and a component X.
The resistance of X changes with temperature, as shown in Figure 2.32.

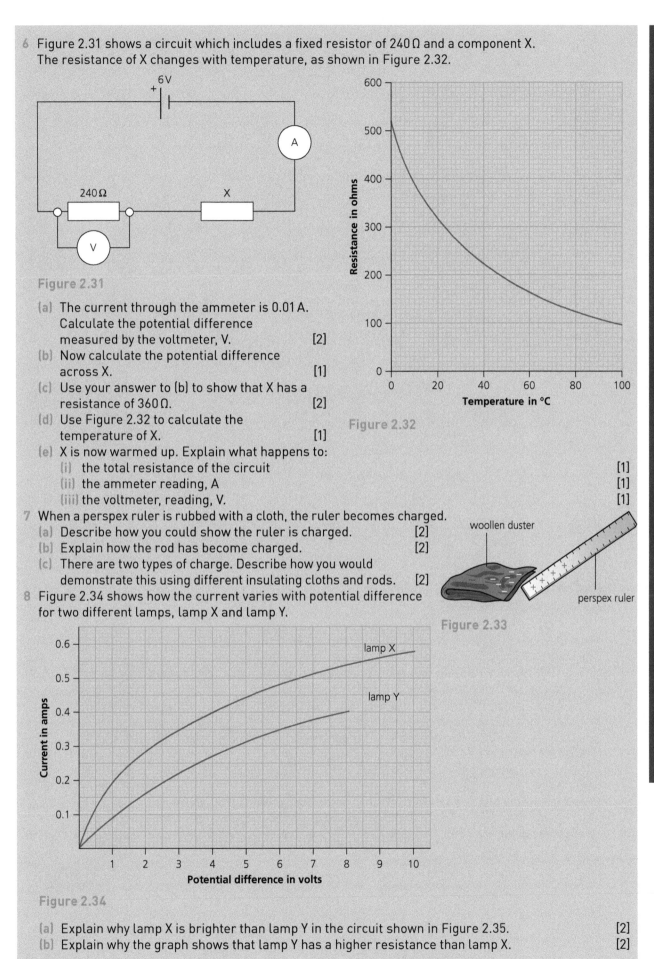

Figure 2.31

(a) The current through the ammeter is 0.01 A.
Calculate the potential difference
measured by the voltmeter, V. [2]
(b) Now calculate the potential difference
across X. [1]
(c) Use your answer to (b) to show that X has a
resistance of 360 Ω. [2]
(d) Use Figure 2.32 to calculate the
temperature of X. [1]

Figure 2.32

(e) X is now warmed up. Explain what happens to:
(i) the total resistance of the circuit [1]
(ii) the ammeter reading, A [1]
(iii) the voltmeter, reading, V. [1]

7 When a perspex ruler is rubbed with a cloth, the ruler becomes charged.
(a) Describe how you could show the ruler is charged. [2]
(b) Explain how the rod has become charged. [2]
(c) There are two types of charge. Describe how you would
demonstrate this using different insulating cloths and rods. [2]

Figure 2.33

8 Figure 2.34 shows how the current varies with potential difference
for two different lamps, lamp X and lamp Y.

Figure 2.34

(a) Explain why lamp X is brighter than lamp Y in the circuit shown in Figure 2.35. [2]
(b) Explain why the graph shows that lamp Y has a higher resistance than lamp X. [2]

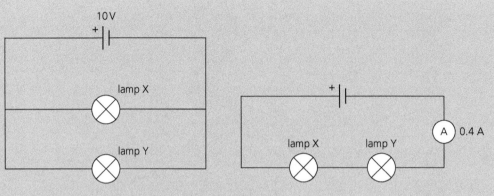

Figure 2.35   Figure 2.36

(c) The lamps are now connected in series as shown in Figure 2.36.
  (i) Use the graph to calculate the potential difference across each bulb now.   [2]
  (ii) Explain which bulb is brighter in this circuit.   [2]

9 A student uses the circuit shown in Figure 2.37 to investigate the way the current through a filament lamp depends on the potential difference across it.
The results of the investigation are shown in the table.

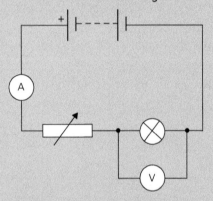

Figure 2.37

Current in amps	0	0.5	0.7	1.2	1.7	2.1	2.3	2.8	3.1
Potential difference in volts	0	0.3	0.8	1.8	3.7	5.0	6.5	9.0	11.0

(a) Which is the dependent variable, current or potential difference?   [1]
(b) Plot a graph of current (y-axis) against potential difference (x-axis). Draw a line of best fit through the points.   [4]
(c) (i) The student made an error in one ammeter measurement. State what the correct reading should have been.
    (ii) Name the type of error that has occurred here and explain what action can be taken to reduce the likelihood of such errors.   [1]
(d) A student extends the line of best fit to predict the current for a potential difference of 12 V. Explain why this prediction is unreliable.   [2]

10 An electric fire element has a power input of 2.3 kW. The metal case of the fire is connected to the earth wire.
(a) Explain why the metal case of the electric fire is connected to the earth wire.   [2]
(b) The charge that flows through the fire element in 10 minutes is 6000 C. Calculate the resistance of the element.   [5]

Answers and quick quiz 2 online

ONLINE

3 Particle model of matter

Density

The density of a material is defined by the equation:

$$\text{density} = \frac{\text{mass}}{\text{volume}}$$

$$\rho = \frac{m}{V}$$

density, ρ, in kilograms per metre cubed, kg/m³
mass, m, in kilograms, kg
volume, V, in metres cubed, m³

Example

Mercury has a density of 13 600 kg/m³. Calculate the mass of 0.002 m³ of mercury.

Answer

$$\rho = \frac{m}{V}$$

So

$$13\,600 = \frac{m}{0.002}$$

$$m = 13\,600 \times 0.002$$

$$= 27.2 \,\text{kg}$$

Solids, liquids and gases

- **Solid.** In a solid, atoms (or molecules) are packed close together in a regular structure. The atoms cannot move from their fixed positions, but they can vibrate. The atoms are held together by strong forces, so it is difficult to change the shape of a solid.
- **Liquid.** The atoms (or molecules) in a liquid are close together. Forces keep the atoms in contact, but the atoms are free to move. A liquid can flow and change shape to fit into any container. Because the atoms are close together, it is difficult to compress a liquid.
- **Gas.** In a gas the atoms (or molecules) are separated by relatively large distances. The forces between the atoms are small. The atoms are in a constant state of random motion. A gas can expand to fill any volume, and a gas is easy to compress.

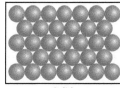
Solid

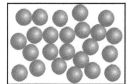
Liquid

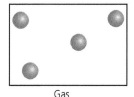
Gas

Figure 3.1

The density of a material can be explained by the particle model of matter:
- Gases have low densities because atoms and molecules are far apart in the gaseous state.
- Metals such as gold are very dense because:
  - the atoms are packed close together
  - each atom has a high mass.

Required practical 5

You should be able to describe how to take appropriate measurements and then calculate the densities of liquids and solids.

The density of a liquid

The cylinder in Figure 3.2 has a mass of 152.6 g when empty and a mass of 169.2 g when it has 20 cm³ of liquid in it. Calculate the density of the liquid in g/cm³.

mass of the liquid = 169.2 − 152.6

$$= 16.6 \text{ g}$$

$$\rho = \frac{m}{V}$$

$$= \frac{16.6}{20}$$

$$= 0.83 \text{ g/cm}^3$$

(a)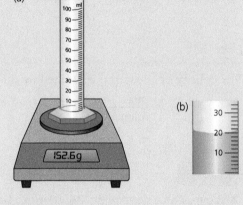

(b)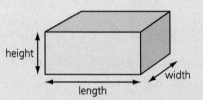

Figure 3.2

The density of a regular solid

A metal block has a height of 5.1 cm, a length of 10.7 cm and width of 9.3 cm. The mass of the block is 3.83 kg. Calculate the density of the metal in kg/m³.

volume of block = 0.051 × 0.107 × 0.093

$$= 5.1 \times 10^{-4} \text{ m}^3$$

$$\rho = \frac{m}{V}$$

$$= \frac{3.83}{5.1 \times 10^{-4} \text{ m}^3}$$

$$= 7500 \text{ kg/m}^3$$

height
length
width

Figure 3.3 The volume of a cuboid = length × width × height.

The density of an irregular solid

Use the information in Figure 3.4 to calculate the density of the rock.

volume of the rock = 20 × 10⁻⁶ m³

$$\rho = \frac{0.09}{20 \times 10^{-6}}$$

$$= 4500 \text{ kg/m}^3$$

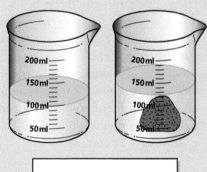

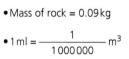

- Mass of rock = 0.09 kg

- $1 \text{ ml} = \dfrac{1}{1\,000\,000} \text{ m}^3$

Figure 3.4

Now test yourself

1 Copy the table and fill in the gaps.

Material	Mass in kg	Volume in m³	Density in kg/m³
A	1800	4.5	
B	0.064		0.08
C		0.01	9000
D	600	0.03	

2 A cube of wood has a side length of 5.7 cm; the mass of the cube is 144.6 g. Calculate the density of the wood in kg/m³. Express your answer to an appropriate number of significant figures.

3 Explain how you would measure the density of an irregularly shaped solid. Include a description of the apparatus you would use and the measurements you would take. Explain what errors might occur in your experiment, and how you would attempt to reduce them.

4 (a) Draw diagrams to show the arrangements of atoms in the three states of matter.
   (b) Use your diagrams to explain why solids are much denser than gases.

Answers on p. 144

Internal energy and energy transfers

Internal energy

Energy is stored inside a system by the particles (atoms or molecules) that make up the system. This is called internal energy. The internal energy is the sum of the kinetic and potential energies of the particles that make up the system.

Heating

Heating increases the energy stored within a system by increasing the internal energy of the particles in the system.

- Heating can increase the temperature of the system – the atoms of the system move faster and the kinetic energy of the atoms rises.
- Heating can cause a change of state – for example, when a liquid evaporates to become a gas. The atoms increase their separation when the substance changes from a liquid to a gas, and this causes the atoms to increase their potential energy. So, the internal energy of the substance increases.

Changes of state

There is an increase in internal energy for:
- melting – a solid turns to a liquid
- boiling or evaporation – a liquid turns to a gas.

There is a decrease in internal energy for:
- freezing – a liquid turns to a solid
- condensation – a gas turns to a liquid.

Changes of state are physical changes. The change does not produce a new substance and the process can be reversed.

Now test yourself

5 (a) Explain what is meant by the term *internal energy*.
   (b) Explain two ways in which heating can increase the internal energy of a substance.
6 (a) What is meant by a *change of state* of a substance?
   (b) Give two examples of a change of state of a substance.

Answers on p. 144

Temperature changes in a system and specific heat capacity

REVISED

If the temperature of a system increases, the increase in temperature depends on the mass of the substance heated, the type of material and the energy supplied to the system.

The following equation applies:

change in thermal energy = mass × specific heat capacity × temperature change

$$\Delta E = mc\Delta\theta$$

The specific heat capacity of a substance is the amount of energy required to raise the temperature of one kilogram of the substance by one degree Celsius.

change in thermal energy, ΔE, in joules, J

mass, m, in kilograms, kg

specific heat capacity, c, in joules per kilogram per degree Celsius, J/kg°C

temperature change, $\Delta\theta$, in degrees Celsius, °C

Example

12000 J of energy are supplied to 4.0 kg of a substance and the temperature of the substances increases by 20°C. Calculate the specific heat capacity of the substance.

Answer

$$\Delta E = mc\Delta\theta$$

$$12\,000 = 4 \times c \times 20$$

$$c = \frac{12\,000}{4 \times 20}$$

$$= 150\,\text{J/kg}°\text{C}$$

Typical mistake

Remember the unit of specific heat capacity is J/kg°C.

Now test yourself

TESTED

7 (a) State the unit of specific heat capacity.
   (b) Explain the meaning of *specific heat capacity*.
8 Use the information in the table to answer the questions.

Substance	Specific heat capacity J/kg°C
Water	4200
Glass	630
Air	1000

(a) A kettle contains 0.5 kg of water. Calculate the energy required to warm the water from 20°C to 70°C.
(b) A glass dish has a mass of 0.2 kg; 12600 J of thermal energy is transferred to the dish. Calculate the temperature rise of the dish.
(c) The mass of air in a room is 75 kg. Calculate the energy required to warm the air from a temperature of −5°C to 20°C.

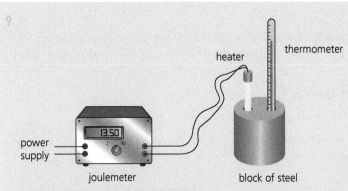

Figure 3.5

(a) The diagram shows the apparatus used to measure the specific heat capacity of steel. The mass of the block is 0.8 kg; the temperature of the block rises from 21 °C to 55 °C. Use the information in the diagram to calculate the specific heat capacity of steel.

(b) Explain why such an experiment is likely to give an answer for the specific heat capacity that is higher than the true value.

Answers on p. 144

Latent heat

REVISED

When a pan of water is put onto a cooker, it heats up and will reach its boiling point at 100 °C. If the pan is left on the cooker, it will continue to boil, but there is no further increase in the water's temperature. Now the energy from the cooker is being used to change the state of the water. The energy increases the internal energy of the steam – as the molecules of water are separated, their potential energy increases. The internal energy of one kilogram of steam at 100 °C is greater than the internal energy of water at 100 °C.

The energy required to change the state of 1 kg of a substance, without a change of temperature, is called the specific latent heat.

> energy for a change of state = mass × specific latent heat

$$E = mL$$

There are three states of matter, so each substance has two specific latent heats:

- The **specific latent heat of fusion** is the energy required to turn 1 kg of a solid into 1 kg of a liquid at the same temperature.
- The **specific latent heat of vaporisation** is the energy required to turn 1 kg of a liquid into 1 kg of a vapour at the same temperature.

Melting and freezing

When a substance melts, energy is supplied to increase the internal energy of the atoms (or molecules). When a substance freezes, the internal energy of the atoms reduces, and energy is released to the surroundings.

Evaporation, condensation and sublimation

When a substance evaporates, energy is supplied to increase the internal energy of the atoms (or molecules). When a substance condenses, the internal energy of the atoms reduces, and energy is released to the surroundings.

> **Exam tip**
>
> When a substance changes from a solid to a liquid, the energy supplied increases the internal energy of the substance, but does not increase the temperature.

> energy, E, in joules, J
>
> mass, m, in kilograms, kg
>
> specific latent heat, L, in joules/kilogram, J/kg

> **Sublimation** is a phase change of a substance directly from a solid to a vapour without passing through a liquid phase.

Example

An immersion heater of power 500W is used to bring a beaker of water to boiling point. The water is allowed to boil for 5 minutes, then the heater is turned off. At the start, the mass of the beaker and water was 632 g, at the end of the mass was 572 g.

Calculate the specific latent heat of vaporisation of water.

Answer

This is an example of a difficult problem and you are expected to use more than one equation. We have to work out E and m first before we use this equation:

$$E = mL$$

To find energy, E:

$$E = \text{power} \times \text{time}$$
$$= 500 \times 5 \times 60$$
$$= 150\,000\,\text{J}$$

To find the mass vaporised, m:

$$m = 632 - 572 = 60\,\text{g} = 0.06\,\text{kg}$$

Now using

$$E = mL$$
$$150\,000 = 0.06 \times L$$
$$L = \frac{150\,000}{0.06}$$
$$= 2\,500\,000\,\text{J/kg or 2.5 MJ/kg}$$

Exam tip

Remember to turn 5 minutes into 300 seconds. Remember to turn 60 g into 0.06 kg.

Exam tip

The unit of specific latent heat is joules/kilogram, J/kg.

Cooling graphs

When a beaker of water cools down, the temperature changes as shown in Figure 3.6. The graph shows a steady drop in temperature, with the rate of cooling slowing down at lower temperatures.

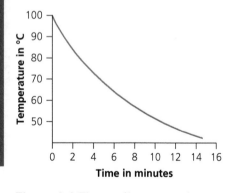

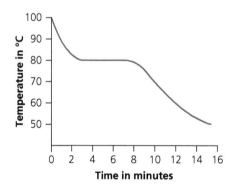

Figure 3.6 **The cooling curve for water.**   Figure 3.7 **The cooling curve for ethanamide.**

When a beaker of ethanamide cools, the cooling curve is very different – see Figure 3.7.

This shows us that ethanamide solidifies (or freezes) at a temperature of 80 °C. The beaker continues to transfer energy to the surroundings, although the temperature stays the same. As the molecules go from the liquid to the solid state, their internal energy decreases, and this allows the energy to be transferred from the ethanamide to the surroundings.

Now test yourself

10 (a) Explain what is meant by the *specific latent heat of fusion* for a solid.
   (b) Give the unit of specific latent heat.
11 (a) Explain why sweating helps us cool down.
   (b) When a dog gets hot it pants. How does panting help the dog to cool?
   (c) A gardener places a large bucket of water in his greenhouse when the weather forecast predicts a severe frost. Explain how the bucket of water can protect the plants from the frost.
12 (a) Figure 3.8 shows a cooling curve for a hot substance.

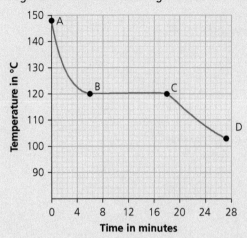

Figure 3.8

   (i) Explain why the substance cools more quickly over the region AB than it does over the region CD.
   (ii) At what temperature does the substance solidify?
   (b) A student writes the following statement.

   'The substance transfers energy to the surroundings at a faster rate at time B than it does at time C.'

   Explain whether this is true.
   (c) Explain whether it is possible to heat a substance without causing the temperature of the substance to increase.
13 A student sets up some apparatus to measure the specific latent heat of fusion of ice.
   The heater has a power of 75 W, and melts 24 g of ice in the 2 minutes.
   (a) Calculate the energy transferred to the ice by the heater in 2 minutes.
   (b) Use the data above to calculate the specific latent heat of fusion of ice. Give your answer in J/kg.
   (c) (i) Explain why it is important that the ice is allowed to reach a temperature of 0 °C, before the heater is switched on.
      (ii) State what type of error is introduced if the ice is warmed up from a temperature of −5 °C.
   (d) Explain what safety precautions you would take in planning this experiment.

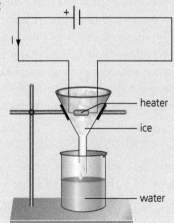

Figure 3.9

Answers on pp. 144–145

The particle model of gases

The particle model helps us to understand the behaviour of gases. The main points of the model are:

- The particles in a gas (atoms or molecules) are in a constant state of random motion.
- The particles in a gas collide with each other and the walls of their container without losing any kinetic energy.
- The temperature of the gas is related to the average kinetic energy of the particles.
- As the average kinetic energy of the molecules increases, the temperature of the gas increases.

Gas pressure

- When the particles of a gas hit the wall of their container, the particles exert a force. In Figure 3.10 each of the molecules exerts a force at right angles to the wall as it bounces off the wall.

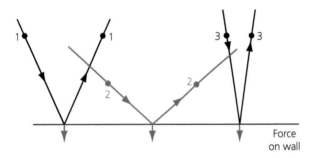

Figure 3.10

- The pressure inside a container of gas, with a fixed volume, increases when the temperature increases. At a higher temperature, the molecules move with greater speed. So they hit the walls of the container harder and more often. The force exerted on the walls increases and the pressure increases.
- The pressure inside a container of gas, at a constant temperature, decreases when the volume is increased. The molecules continue to move with the same average kinetic energy (and therefore speed), but the particles hit the walls less frequently. So the force exerted on the wall and, therefore, the pressure decreases.

Expanding and compressing gases

Figure 3.11 shows apparatus that can be used to change the pressure and volume of a fixed mass of air at a constant temperature.

Figure 3.12 shows the relationship between the pressure and the volume of air in the column of gas in Figure 3.11.

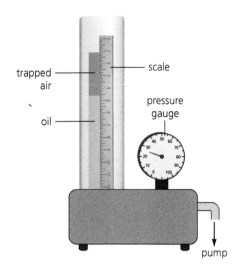

Figure 3.11

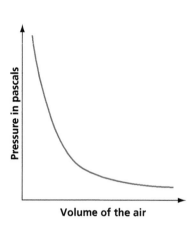

Figure 3.12

By increasing the pressure on the oil, using a pump, the column of trapped air is compressed.

For a fixed mass of gas held at constant temperature:

pressure × volume = constant

$$PV = \text{constant}$$

pressure, P, in pascals, Pa

volume, V, in metres cubed, m³

There is an inverse proportion between the volume and the pressure of a gas.

$$P \propto \frac{1}{V}$$

For example:
- halving the volume of the gas doubles the pressure
- increasing the volume of the gas by a factor of three, reduces the pressure to a third of the original value.

Example

A gas in a container of volume 2.0 m³ has a pressure of 500 kPa. The volume is reduced to 0.40 m³. Calculate the new pressure.

Answer

$$P_1V_1 = P_2V_2$$

$$500 \times 2 = P_2 \times 0.4$$

$$P_2 = \frac{500 \times 2}{0.4}$$

$$= 2500 \,\text{kPa}$$

⊕ Work and energy

REVISED

When you pump up your bicycle tyres, you will find that the pump gets hot. This can be explained in two ways:
- The moving piston collides with air molecules and the molecules bounce off the piston at a greater speed. The average kinetic energy of the molecules increases, so the temperature of the gas increases.
- We can also explain the temperature rise using the idea of work. Work is the transfer of energy by a force. So when you push the pump you do work. The work increases the internal energy of the gas, and the temperature of the gas rises.

work done = force × distance moved in the direction of the force

Now test yourself

14 (a) Describe the motion of the particles in a gas.
   (b) How does the motion of the particles change when the gas gets hotter?
15 Use the particle model of gas to explain:
   (a) how a gas exerts a pressure on the walls of its container
   (b) why the pressure of a gas, which is kept at constant volume, increases when its temperature is increased.
16 A balloon has a volume of $0.02\,m^3$ when the pressure inside it is $100\,kPa$. A weight is attached to the balloon and it falls to the bottom of a lake where the pressure is $500\,kPa$. Calculate the volume of the balloon now. Assume the temperature of the balloon doesn't change.
17 A cylinder of gas has a volume of $0.40\,m^3$ at a pressure of $700\,kPa$. Calculate the volume of the gas when the valve on the cylinder is opened and the gas is allowed to escape and reach atmospheric pressure of $100\,kPa$. Assume the temperature of the gas doesn't change.

Answers on p. 145

Summary

- Density is measured in kg/m^3.

  density = mass/volume

- There are three states of matter: solid, liquid and gas.
  - The particles in a solid are closely packed and vibrate about fixed positions.
  - The particles in a liquid are in close contact, but are free to move past each other, so a liquid flows.
  - The particles in a gas are far apart, and they move randomly in all directions.
- The internal energy of a system of particles is the sum of the kinetic energy and potential energy of the particles. Heating increases the internal energy of the particles.
- Heating can raise the temperature of a substance. The particles move faster as the temperature rises.
- Heating can also cause a change of state, without changing the temperature of a substance.
- When heating causes a change of temperature the following equation applies:

  change in thermal energy = mass × specific heat capacity × temperature change

  $$\Delta E = mc\Delta\theta$$

  c is measured in $J/kg\,°C$

- When heating causes a change of state, the following equation applies.

  energy for a change of state = mass × specific latent heat

  $$E = mL$$

  L is measured in J/kg
- When a liquid cools, its temperature remains constant as it solidifies (see Figure 3.7).
- The particle model of gases states that gas molecules (or atoms) are in a constant state of random motion. Particles hit the walls of their container and exert a pressure. The pressure increases at higher temperatures; because the particles have greater kinetic energy, they hit the walls faster and more often.
- For a gas at constant temperature:

  pressure × volume = constant

  $$PV = \text{constant}$$

- The temperature of a gas can be increased by doing mechanical work to compress it.

Exam practice

1 Which of the following is the correct unit for specific heat capacity?
   - A   J/kg
   - B   J kg °C
   - C   J/kg °C
   - D   J   [1]

2 The diagrams show the arrangements of particles in a solid and a gas.

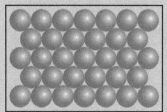 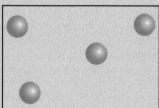

Figure 3.13

   (a) (i)   Describe the motion of the particles in the solid at room temperature.   [1]
      (ii) Explain what happens to the particles when the temperature rises.   [1]
   (b) (i)   Describe the motion of the particles in the gas.   [1]
      (ii) Explain why the particles in a gas exert a pressure on the walls of their container.   [2]
      (iii) Explain why the pressure exerted by the gas increases when the temperature of the
            gas rises.   [2]
   (c) A mixture of ice and water is put into a pan at a temperature of 0 °C. The pan is heated until
       all the ice melts.
      (i)   Draw a diagram to show the arrangement of particles in a liquid.   [1]
      (ii) Explain why the temperature of the ice and water remains at 0 °C until all the ice is
            melted, even though the mixture is being heated.   [3]

3 A student uses a ruler to measure the side lengths of a cuboid of aluminium as shown in
  Figure 3.14. Aluminium has a density of 2700 kg/m³.

   (a) Calculate the volume of the aluminium in m³.   [2]
   (b) Use the density of aluminium to calculate its mass.   [2]
   (c) When the student puts the aluminium onto an electronic
       balance, he records a mass of 205.8 g. Explain what might
       have caused the difference between the actual and
       calculated value of the mass.   [2]

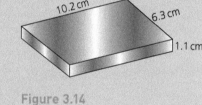

Figure 3.14

4 A heater is used to heat a metal block of mass 0.5 kg. After the
  heater is turned on the temperature rises, as shown in Figure 3.15.

   (a) Use the graph to determine the temperature rise in the first 60 s of heating.   [1]
   (b) During the first 60 s of heating 5025 J of energy is supplied to the block. Calculate the specific
       heat capacity of the block.   [3]
   (c) Use the information in part (b) to calculate the power of the heater.   [2]

5 A small heater is placed into some crushed ice, and turned on for 4 minutes. The heater has
  a power of 36 W.
   (a) Show that the heater transfers 8600 J to the ice in
       4 minutes.   [2]
   (b) The specific latent heat of fusion of ice is 330 kJ/kg.
       Calculate the mass of ice melted after 4 minutes.   [3]

6 Some air is trapped in an airtight cylinder. A piston slowly
  compressed the air, so that the length of the air column is
  reduced from 40 cm to 15 cm. The air temperature remains
  constant. The initial air pressure, at a length of 40 cm, is 120 kPa.
   (a) Explain in terms of the particle model of gases why the pressure
       of the air increases as the air column is compressed.   [2]

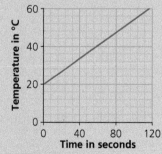

Figure 3.15

(b) (i) Calculate the air pressure when the column has a length of 15 cm. [3]

(ii) The cross-sectional area of the cylinder is 0.06 m^2 and the mass of gas in the cylinder is 0.02 kg. Calculate the density of the gas when the length of the air column is 15 cm. Express your answer in kg/m^3. [3]

7 The air in a room is to be heated from 5 °C to 18 °C by a convector heating. Use the information in the list below to calculate the energy required to heat the air. [6]

- Volume of air in the room: 70 m^3
- Density of air: 1.2 kg/m^3
- Specific heat capacity of air: 1000 J/kg °C

Answers and quick quiz 3 online

ONLINE

4 Atomic structure

The structure of an atom

Atoms are very small, having a radius of about 1×10^{-10} m.

An atom has a nucleus which has a radius less than $1/10\,000$ of the atom.

The nucleus contains positively charged protons and neutral neutrons.

Negatively charged electrons are arranged at different distances from the nucleus, which correspond to different energy levels. Electrons can change energy levels by the absorption or emission of electromagnetic radiation.

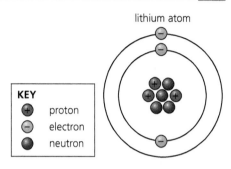

lithium atom

KEY
- ⊕ proton
- ⊖ electron
- ⬤ neutron

Figure 4.1 **The arrangement of protons, neutrons and electrons in a lithium atom. Note this is not drawn to scale.**

Atoms and ions

In an atom the number of electrons is equal to the number of protons. Atoms have no overall charge, because the size of the negative charge on an electron is the same size as the positive charge on a proton.

If an atom gains an electron, it becomes a negative ion. If the atom loses an electron, it becomes a positive ion.

Mass number, atomic number and isotopes

An atom is determined by the number of protons in its nucleus. The number of protons in an element is called its **atomic number**.

The total number of protons and neutrons in an atom is called its **mass number**.

An atom can be represented as shown in this example:

mass number 27
atomic number 13 Al

This symbol tells us that aluminium has 13 protons in its nucleus and 14 neutrons, making a total of 27 protons and neutrons.

Isotopes

Not all the atoms of an element have the same mass, for example one atom of aluminium might have a mass of 27 and another a mass of 26. Both atoms have 13 protons, but one has 13 neutrons and the other 14 neutrons. These are two **isotopes** of aluminium:

- $^{27}_{13}\text{Al}$ is the most common isotope of aluminium.
- $^{26}_{13}\text{Al}$ is another isotope of aluminium.

> **Atomic number** is the number of protons.
>
> **Mass number** is the number of protons and neutrons.
>
> **Isotopes** are different forms of a particular element. Isotopes have the same number of protons but different numbers of neutrons.

Now test yourself

1 (a) Which of the following is the approximate radius of an atom?

   10^{-4} m 10^{-7} m 10^{-10} m

   (b) Use one of the answers from the list below to complete the following sentence.

   The radius of an atom is approximately .... times the radius of a nucleus.

   100   10 000   10 000 000

2 An oxygen atom has 8 protons, 8 neutrons and 8 electrons.
   (a) State the atomic number of oxygen.
   (b) State the mass number of oxygen.
   (c) Explain why an oxygen atom is neutral.

3 Calculate the number of protons and neutrons in each of the following nuclei:
   (a) $^{11}_{5}\text{B}$
   (b) $^{32}_{16}\text{S}$
   (c) $^{156}_{64}\text{Gd}$
   (d) $^{237}_{93}\text{Np}$

4 Explain the meaning of the following terms:
   (a) atomic number
   (b) mass number
   (c) isotope
   (d) ion

5 The atomic radius of uranium is about 1.8×10^{-10} m and the nuclear radius of uranium is about 7.5×10^{-15} m. Calculate the ratio of atomic to nuclear radii in uranium.

Answers on p. 145

The development of the model of the atom

New experimental evidence may lead to a scientific model being changed.

Until the discovery of the electron in 1897, atoms were thought to be indivisible solid spheres, and the smallest part of matter.

The 'plum pudding' model

In 1904, J J Thompson proposed a new model for the atom. The idea of the model was that an atom was made up of a positive ball of matter, with electrons dotted inside – the electrons are the plums in the pudding.

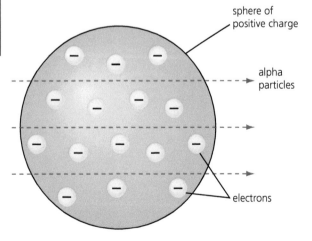

sphere of positive charge

alpha particles

electrons

Figure 4.2 The 'plum pudding' atomic model.

Exam practice answers and quick quizzes at **www.hoddereducation.co.uk/myrevisionnotes**

The nuclear model of the atom

In 1909 the Geiger and Marsden experiment led to the idea of the nuclear atom. They directed a beam of alpha particles (He^{2+} nuclei) at a thin gold foil.

They expected the alpha particles to travel straight through as shown in Figure 4.2.

In fact, most alpha particles did travel straight through the foil. But a very small fraction of alpha particles bounced back, as shown in Figure 4.3.

The conclusion we now draw from the Geiger and Marsden experiment is that the alpha particles are repelled by a very small, positively charged nucleus, which contains most of the mass of the atom.

- The nucleus must be small because only a small fraction of alpha particles bounce back.
- The nucleus is positive because its strong electric field repels the positively charged alpha particles.
- The nucleus must be massive, because a small nucleus would be knocked forwards by the alpha particle.

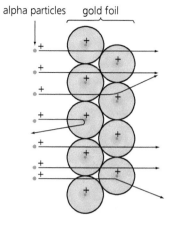

alpha particles   gold foil

Figure 4.3

The Bohr model of the atom

In 1913 Neils Bohr suggested a model of the atom in which electrons move round the nucleus in circular orbits. In this model electrons can change their orbit.

Scientists had discovered that matter absorbs and emits specific energies or electromagnetic radiation.

The Bohr model explains this:

- When an electron falls from a high level to a low level, it emits electromagnetic radiation – for example, falling from level 3 to 2 in Figure 4.4.
- An electron jumps up a level by absorbing electromagnetic radiation.

Later experiments showed that the nucleus could be divided further into protons and neutrons. James Chadwick discovered neutrons in 1932.

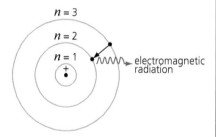

Figure 4.4

Now test yourself

TESTED

6 Explain and describe the experimental evidence that led to the nuclear model of the atom.
7 Describe the Bohr model of the atom.

Answers on p. 145

Atoms and nuclear radiation

REVISED

Some atomic nuclei are unstable. The nucleus gives out radiation and it changes to become more stable. This is a random process which is called radioactive decay.

The activity of a radioactive source is the rate at which it decays.

Activity is measured in Becquerel, Bq.

1 Bq = 1 nuclear decay per second.

A small radioactive source might have a decay rate of 10^6 Bq.

Ionising radioactive particles may be detected using a Geiger–Müller (GM) tube.

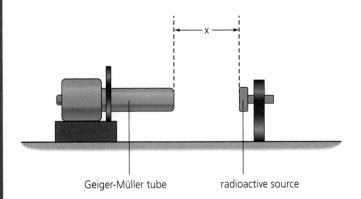

Figure 4.5

The count rate detected by a GM tube is always less than the activity of a radioactive source, because the source emits particles in all directions.

Nuclear radiation

Nuclear radiation that may be emitted from nuclei include:
- an alpha particle (α) – this consists of two protons and two neutrons, which is the same as a helium nucleus
- a beta particle (β) – this is a high-speed electron that escapes from a nucleus when a neutron turns into a proton
- a gamma (γ) ray – this is electromagnetic radiation emitted from the nucleus
- a neutron (n).

Properties of radiation

- Alpha particles travel about 5 cm through air and can be stopped by a piece of paper. Alpha particles are strongly ionising.
- Beta particles can travel several metres through air and can be stopped by a sheet of aluminium that is a few millimetres thick. Beta particles are not as strongly ionising as alpha particles.
- Gamma rays can only be effectively stopped by very thick lead. Gamma rays are only weakly ionising and travel great distances in air.

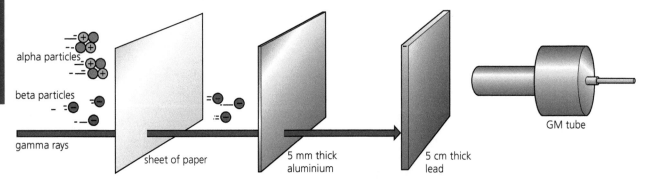

Figure 4.6

Radiation	Nature	Range in air	Ionising power	Penetrating power
Alpha α	helium nucleus	a few centimetres	very strong	stopped by paper
Beta β	electron	a few metres	medium	stopped by aluminium
Gamma γ	electromagnetic waves	great distances	weak	stopped by thick lead

Radiation damage

Radiation that gets into our bodies can cause damage to our cells. Alpha particles cause the most damage – this could happen if we inhaled a radioactive gas.

Gamma rays are less ionising than alpha particles, but they can get into our body because they are very penetrating.

Nuclear equations

An alpha particle may be represented by the symbol:

$${}^{4}_{2}\text{He}$$

So when an alpha particle is emitted from a nucleus, it causes the mass number to decrease by 4 and the atomic number by 2.

For example:

$${}^{225}_{89}\text{Ac} \rightarrow {}^{221}_{87}\text{Fr} + {}^{4}_{2}\text{He}$$

A beta particle may be represented by the symbol:

$${}^{0}_{-1}\text{e}$$

So when a beta particle is emitted from a nucleus, the mass number remains the same but the atomic number increases by 1.

For example:

$${}^{3}_{1}\text{H} \rightarrow {}^{3}_{2}\text{He} + {}^{0}_{-1}\text{e}$$

The emission of a gamma ray from a nucleus does not cause the mass or atomic number to change. The gamma ray has no mass or charge, but it does carry away some energy from the nuclear store.

Typical mistake

Remember that when a beta particle is emitted, the atomic number **increases** by 1; it does **not** decrease.

Now test yourself

TESTED

8   Which of the following are properties of beta radiation?
    A   It is the most strongly ionising radiation.
    B   It is stopped by aluminium.
    C   It is a fast-moving electron.
9   Explain why a teacher uses long tongs when she handles radioactive sources.
10 (a) Explain what is meant by the *activity* of a radioactive source.
    (b) Which of the following is the correct unit for the activity of a radioactive source?
         rutherford   geiger   becquerel
11 Explain the nature of each of the following:
    (a) an alpha particle
    (b) a beta particle
    (c) a gamma ray.
12 When a gamma ray is emitted from a nucleus, what changes occur to the mass and atomic numbers of the nucleus?
13 Fill in the gaps in the following radioactive equations:
    (a) ${}^{241}_{94}\text{Pu} \rightarrow {}^{?}_{92}\text{U} + {}^{4}_{?}\text{He}$
    (b) ${}^{237}_{92}\text{U} \rightarrow {}^{237}_{?}\text{Np} + {}^{0}_{-1}\text{e}$
    (c) ${}^{?}_{26}\text{Fe} \rightarrow {}^{59}_{27}\text{Co} + {}^{?}_{?}\text{e}$
    (d) ${}^{213}_{84}\text{Po} \rightarrow {}^{?}_{?}\text{Pb} + {}^{4}_{2}\text{He}$
    (e) ${}^{32}_{14}\text{Si} \rightarrow {}^{32}_{15}\text{P} + ?$
    (f) ${}^{229}_{90}\text{Th} \rightarrow {}^{225}_{88}\text{Ra} + ?$

Answers on pp. 145–146

Half-lives and the random nature of radioactive decay

Radioactive decay is random.

We use the analogy of rolling lots of dice to help explain this. If you roll one die, it is not possible to predict if you will throw a six. If you throw 600 dice, on average you would expect to throw about 100 sixes. Radioactive decay is like that: you cannot predict when one nucleus will decay, but when there are lots of nuclei they decay in a predictable way.

Half-life

The half-life of a radioactive isotope is the time it takes for the number of nuclei in a sample to halve. The half-life is also the time it takes for the count rate (or activity) detected by a GM tube (or other detector) to halve.

The half-life is a measure of the stability of a nucleus. A shorter half-life means the nucleus is less stable.

Measuring half-lives

You can determine the half-life by measuring the activity of a radioactive isotope. The graph shows how the measured count rate for an isotope changes with time. You can see that the half-life for this isotope is 50 seconds. Every 50 seconds the count rate halves.

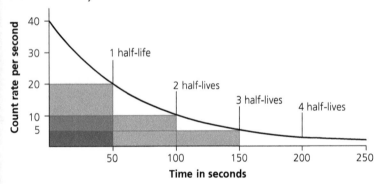

Figure 4.7

Example

A radioactive isotope has an activity of 1.6×10^6 Bq. The half-life of the isotope is 8 hours. Calculate the activity of the isotope after a day.

Answer

One day is 24 hours which is three half-lives.

So the activity will be

$$\frac{1}{2} \times \frac{1}{2} \times \frac{1}{2} \times 1.6 \times 10^6 \, \text{Bq}$$

$$= 2 \times 10^5 \, \text{Bq}$$

Exam tip

The activity of a radioactive isotope is:

- $\frac{1}{2}$ after 1 half-life
- $\frac{1}{4}$ after 2 half-lives
- $\frac{1}{8}$ after 3 half-lives
- $\frac{1}{16}$ after 4 half-lives.

Now test yourself

14 Use a word from the list to complete the sentence.

count rate   activity   reaction

The .... is the number of particles emitted in one second by a radioactive source.

15 Explain what the word *random* means.

16 The graph Figure 4.8 shows the decay of three different radioactive isotopes.
Which isotope:
   (a) has the longest half-life
   (b) has the shortest half-life
   (c) could be used as a medical tracer?

17 A scientist measures the count rate for a radioactive isotope. His measurements are shown in the table.

Count rate Bq	Time in hours
12 000	0
9240	2
7110	4
5480	6
4220	8
3250	10
2500	12

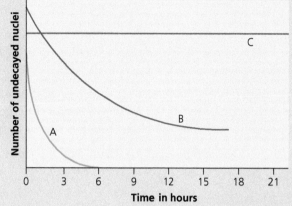

Figure 4.8

Plot a graph to determine the half-life of the isotope.

18 An isotope has a half-life of 16 days.
Today its activity is measured to be 4.0×10^5 Bq.
Calculate its activity in 64 days' time.

Answers on p. 146

Radioactive contamination

There is an important difference between **irradiation** and radioactive **contamination**.

● A patient may be exposed to radiation in a course of radiotherapy. Then the body is irradiated by a specific dose of radiation. Once the radioactive source is removed, the patient's body is not radioactive.

● Radioactive contamination occurs when unwanted radioactive material is absorbed by another material. For example, if there is a leak of radioactive waste from a power station, radioactive material may flow into a river. Then animals that drink from the river absorb radioactive materials into their bodies. In this way an animal is contaminated and continuously exposed to radiation.

> **Exam tip**
>
> Make sure you know the difference between irradiation and contamination.

Screening from radiation

When a radioactive source is used to irradiate something, the operator takes precautions to avoid exposure to radiation. For example, radiographers in hospitals wear lead aprons, and keep well away from a source when it is being used.

Radioactive contamination provides a great risk to us, as we cannot shield ourselves – if we have absorbed a contaminated material, the radioactive source is inside our body.

Hazards and uses of radioactive emissions

Background radiation

We are exposed to a low level of radiation due to radioactive sources in the environment. The main sources of background radiation are:
- rocks that produce radon gas
- cosmic rays from space
- buildings and the ground
- food and drink.

Many rocks contain radioactive isotopes and some rocks produce radon gas. This is particularly dangerous as radon emits alpha particles which can get into our lungs if we inhale the gas.

Radioactive isotopes also get into our food and drink, or into the materials which are used to build our houses. Energetic radioactive emissions (cosmic rays) reach us from the Sun or deep space.

In this way we are exposed to radiation all the time.

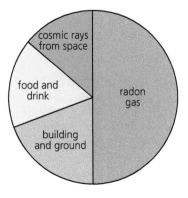

Figure 4.9 **Natural sources of background radiation in Britain.**

Man-made radiation

We are also exposed to radiation due to man-made sources. For example, in 1986 the background radiation in Britain increased due to radioactive emissions from the Chernobyl reactor in Ukraine.

People are sometimes exposed to radiation when they undergo medical treatment, or some workers are exposed to radiation at their place of work.

> **Exam tip**
>
> Background radiation comes from natural and man-made sources in our environment.

Radiation dose

The radiation dose that our bodies receive is measured in sieverts, Sv, or millisieverts, mSv. The dose is a measure of the energy imported in our body and the damage done by a particular radiation.

In Britain, the typical background dose per year is 2 mSv. The background dose we receive varies depending on our location and/or occupation.

The maximum dose allowed per year for a worker in an industrial setting is 50 mSv.

Uses of radiation

Nuclear radiations are used in medicine mainly in two ways:
- for the examination of internal organs
- for the destruction or control of unwanted tissues.

Tracers

Technetium-99 is a commonly used **tracer**. This radioactive isotope is attached to a biochemical agent which is then absorbed by the organ to be examined.

Technetium-99 has many advantages:
- It emits gamma rays that can be detected outside the body.
- It has a half-life of 6 hours.

A short half-life means the isotope has a high activity, so the gamma rays are easily detected for a short time. But as soon as the examination has finished, the short half-life ensures the isotope is not emitting gamma rays for long inside the body.

> A **tracer** is a radioactive isotope that is absorbed by the body. The radioactive emissions allow doctors to monitor the body.

Tissue destruction

Some cancers can be destroyed with high doses of radiation. Sometimes gamma rays are directed from outside the body. Other cancers are treated by chemicals inside the body that emit alpha or beta radiation – short range radiation supplies a dose directly to the cancer.

Although exposure to radiation carries a risk, this is a smaller risk than leaving a cancer untreated.

Half-life and radiation risk

Radioactive isotopes have a wide variety of half-lives. The hazard associated with radioactive material depends on the half-life.

- **Medical tracers**. In this situation, doctors want to use a tracer with a short half-life so that it has a high activity for a very short time. This reduces the dose received by a patient.
- **Nuclear waste**.
  - Isotopes with a short half-life are highly active, but for a short time. For example, an isotope with a half-life of 8 days is a hazard for a few months. However, after a few years there is little risk.
  - Isotopes with a long half-life provide a risk for many years. For example, caesium-137 has a half-life of 30 years. Therefore, an area contaminated by caesium-137 might be a hazard for hundreds of years.

Now test yourself

TESTED

19 (a) Explain what is meant by the term *background radiation*.
   (b) Explain why background radiation varies depending on the location.
20 Explain why an isotope used as a medical tracer should:
   (a) have a short half-life
   (b) be an emitter of gamma rays.
21 Explain the difference between irradiation and contamination.
22 In the UK the average annual background dose of radiation is 2 mSv.
   Two patients, A and B, are exposed to radiation during medical treatment.
   - Patient A receives a dose of 10 mSv from a tracer during an investigation of a kidney.
   - Patient B receives a dose of 700 mSv to treat a cancer.
   Evaluate the risks and benefits to each patient.
23 In a nuclear accident two radioactive isotopes, X and Y, contaminate a power station. Figure 4.10 shows how the activities of the isotopes change with time. Discuss the hazard caused by each isotope:
   (a) in the first year
   (b) after 40 years.

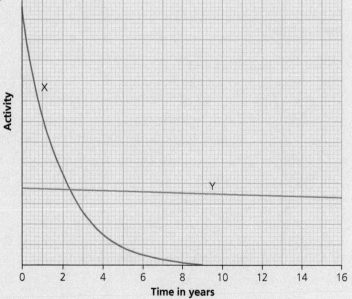

Answers on p. 146

Figure 4.10

Nuclear fission

Nuclear fission is the splitting of a large nucleus – for example, uranium or plutonium.

Sometimes a nucleus spontaneously splits into two, but this is rare. Fission is usually triggered by a neutron – see Figure 4.11.

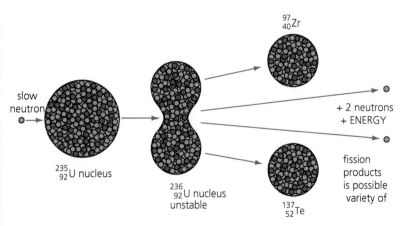

Figure 4.11

A nucleus splits into two smaller nuclei of roughly the same size. Two or three further neutrons and some gamma rays are emitted.

All the products of fission have a store of kinetic energy.

Chain reaction

Neutrons that are emitted during fission can strike other neighbouring nuclei and cause them to split too. This can produce a chain reaction as shown in Figure 4.12.
- A controlled chain reaction is used in a nuclear power station. The energy from the fission products can be used to generate electricity.
- In an uncontrolled chain reaction, the energy is produced at such a rate that a nuclear explosion is produced. This is the principle behind a nuclear bomb.

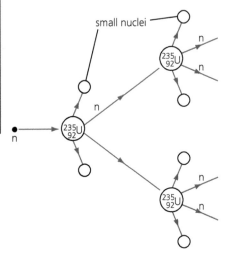

Figure 4.12

Nuclear equations

Fission can be described by nuclear equations. For example:

$$^{1}_{0}n + ^{235}_{92}U \rightarrow ^{137}_{52}Te + ^{97}_{40}Zr + 2^{1}_{0}n + \text{gamma rays}$$

Nuclear fusion

Two small nuclei can fuse together at very high temperatures to form a heavier nucleus. This process happens in stars. Some of the mass of the smaller nuclei is turned into the energy of radiation.

For example, this reaction can occur in the Sun:

$$^{2}_{1}H + ^{3}_{1}H \rightarrow ^{4}_{2}He + ^{1}_{0}n + \text{energy}$$

Now test yourself

TESTED

24 Name the particle that causes the fission of a nucleus of plutonium-239.
25 Name the process by which two small nuclei join together to form a larger nucleus.
26 (a) Explain what is meant by *nuclear fission*.
   (b) Explain how nuclear fission can be continued by a chain reaction.
27 Copy and complete the following nuclear equation which shows the fission of plutonium.

$$^{1}_{0}n + ^{239}_{94}Pu \rightarrow ^{134}_{?}Xe + ^{?}_{40}Zr + 3^{1}_{0}n$$

Answers on p. 146

Typical mistake

When balancing an equation involving fission, remember to include the neutron on the left-hand side.

Summary

- An atom has a radius of about 10^{-10} m.
- The radius of a nucleus is less than 1/10 000 of the radius of an atom.
- A nucleus contains positively charged protons and neutral neutrons. A neutral atom contains as many negatively charges electrons as protons.
- The atomic number of an atom is the number of protons.
- The mass number of an atom is the sum of the numbers of protons and neutrons.
- These numbers may be represented:

  $$\text{mass number 16} \atop \text{atomic number 8}}O$$

- An element is determined by the number of protons in the nucleus. Different isotopes of an element have different numbers of neutrons.
- The 'plum pudding' model of the atom was an early model which suggested the atom was a solid, positively charged mass with electrons inside.

- The scattering of alpha particles led to the nuclear model of the atom, with a small, massive, positively charged nucleus surrounded by electrons.
- The Bohr model of the atom stated that electrons orbit the nucleus and can change their energy level by absorbing or emitting electromagnetic radiation.
- An alpha particle is a helium nucleus. It is strongly ionising, travels about 5 cm in air and is stopped by paper. It is represented by:

  $$^{4}_{2}He$$

- A beta particle is a fast electron. It is less ionising than an alpha particle and is stopped by aluminium a few millimetres thick. It is represented by:

  $$^{0}_{-1}e$$

- A gamma ray is an electromagnetic ray. It is weakly ionising and only stopped by very thick lead.
- Alpha emission reduces the mass number of a nucleus by 4 and the atomic number by 2. For example:

$$^{219}_{86}\text{Rn} \rightarrow {}^{215}_{84}\text{Po} + {}^{4}_{2}\text{He}$$

- Beta emission increases the atomic number of a nucleus by 1 and leaves the mass number unchanged. For example:

$$^{14}_{6}\text{C} \rightarrow {}^{14}_{7}\text{N} + {}^{0}_{-1}\text{e}$$

- Radioactive decay is random.
- Radioactive decay has a half-life. For every half-life that passes, the activity of a radioactive source and the number of radioactive nuclei halves.

- The activity of a radioactive source is the number of emissions per second. This is measured in becquerel, Bq.
- Radioactive contamination occurs when unwanted radioactive materials are present in other materials. Irradiation occurs when a material is exposed to an external source of radiation.
- Make sure you know about the hazards and uses of radiation.
- Nuclear fission occurs when a large nucleus absorbs a neutron and splits into two smaller nuclei.
- Neutrons emitted in nuclear fission can maintain a chain reaction.
- Nuclear fusion is the joining together of two smaller nuclei to make a larger nucleus.

Exam practice

1 A radioactive source emits alpha (α), beta (β) and gamma (γ) radiation.
  (a) Which of the radiations is most ionising? [1]
  (b) Which two radiations will pass through a piece of paper? [1]
  (c) Which radiation has the greatest range in air? [1]

2 (a) Fresh raspberries are sometimes irradiated before being transported to the UK. The irradiation kills bacteria on the raspberries.
    Which of these statements is true?
      A The raspberries become radioactive and cannot be eaten for a week.
      B The irradiation contaminates the raspberries.
      C Radioactive particles settle on the raspberries.
      D The raspberries do not become radioactive, and are safe to eat. [1]
  (b) Suggest a reason why a farmer would want to irradiate his raspberries. [1]

3 Figure 4.13 represents an atom of boron-11.
  (a) State the number of
    (i) protons
    (ii) neutrons
    (iii) electrons in the atom. [3]
  (b) State the atomic number of boron. Give a reason for your answer. [2]
  (c) Boron-12 is a radioactive isotope of boron.
    (i) Explain the word *isotope*. [1]
    (ii) How does boron-12 differ from boron-11? [1]
  (d) Boron-12 decays by the emission of a beta particle.
    (i) Which of the following describes a beta particle?
      A a helium nucleus
      B an electron from the nucleus
      C an electromagnetic wave [1]
    (ii) Complete the following equation that describes the decay of boron-12. [2]

$$^{12}_{5}\text{B} \rightarrow {}^{?}_{?}\text{C} + {}^{?}_{?}\text{beta}$$

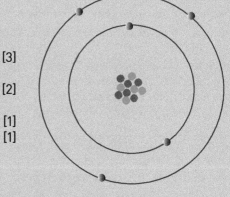

Figure 4.13

→

4 Gadolinium-148, $^{148}_{64}$Gd, is a radioactive isotope that decays by emitting alpha particles.

(a) (i) State the atomic number of gadolinium [1]

(ii) Calculate the number of neutrons in a nucleus of gadolinium-148. [1]

(b) Complete the following equation that describes the decay of gadolinium-148. [2]

$$^{148}_{64}\text{Gd} \rightarrow\ ^{?}_{62}\text{Sm} +\ ^{4}_{?}\text{He}$$

(c) The graph shows how the activity of a sample of gadolinium-148 changes over a period of time.

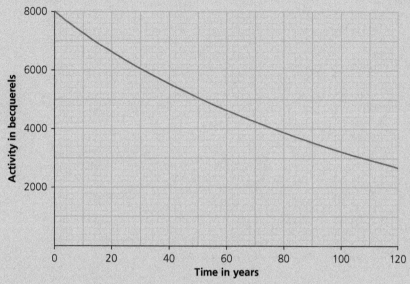

Figure 4.14

(i) Explain what is meant by the word *activity*. [1]

(ii) Calculate the half-life of gadolinium-148. [2]

(iii) Use your answer to (ii) to predict how long it will take for the activity to drop from 8000 Bq to 1000 Bq. [2]

5 The background count in Britain is about 2 mSv per year.

(a) State two natural sources of radiation that contribute to the background count. [2]

(b) The table gives information about different radiation doses.

Radiation dose in mSv	Information
10 000	Likely to cause death within days
1000	A targeted dose to treat cancer
100	The lowest dose per year to increase the risk of cancer
10	The dose from a computer tomography (CT) scan

Discuss the risks and likely benefits to a patient of:

(i) a 1000 mSv targeted dose to treat cancer [2]

(ii) a CT scan. [2]

(c) The table shows information about three radioactive isotopes. Explain with reasons which istotope could be used safely as a medical tracer. [3]

Isotope	Radiation emitted	Half-life
A	alpha	4 minutes
B	gamma	100 days
C	gamma	6 hours

6 Figure 4.15 shows what can happen to a nucleus of uranium when it absorbs a neutron.

(a) What is the name given to this process? [1]

(b) Copy and complete the diagram to show how this process could lead to a chain reaction. [3]

7 In the early twentieth century, scientists thought that atoms were made up of electrons embedded into a ball of positive charge (Figure 4.16).

New evidence led to a new model of the atom. Explain what that evidence was and how it led to a different atomic model. [6]

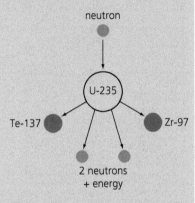

Figure 4.15

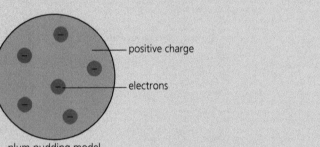

positive charge

electrons

plum pudding model

Figure 4.16

8 Explain the advantages and disadvantages of generating electricity using nuclear power.

Answers and quick quiz 4 online

ONLINE

5A Forces

Forces and their interactions

Scalars and vectors

Scalar quantities have magnitude (size) only. Examples include:
- speed
- mass
- distance
- energy.

Vector quantities have both magnitude (size) and direction. Examples include:
- force
- acceleration
- velocity.

We represent vectors with an arrow; the direction of the arrow shows the direction of the vector and the length of the arrow the magnitude of the quantity.

> **Exam tip**
>
> Velocity and speed are often used to mean the same thing. But speed is a scalar (e.g. 15 m/s) and velocity is a vector, which has both magnitude and direction (e.g. 15 m/s to the right).

Now test yourself

TESTED ☐

1 Which of the following quantities are vector quantities?

force   mass   distance   acceleration
speed   energy   velocity

2 What is wrong with the following statement?

'A force of 3 N acts on an object.'

3 Car A travels due north on a motorway at 30 m/s. Car B travels due south on the motorway at 15 m/s.

Draw vectors to represent these two velocities.

Answers on p. 146

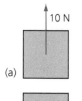
(a)

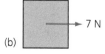
(b)

(c)

Figure 5.1 **Examples of vectors.**

Contact and non-contact forces

- Contact forces act when one body touches another. Examples of contact forces include: friction, air resistance, tension in a rope, and the normal contact force when one object rests against another.
- When non-contact forces act, bodies are physically separated. Gravitational, electrostatic and magnetic forces are non-contact forces.

Gravity

Weight is the force that acts on an object due to gravity.

The force of gravity around the Earth is due to the gravitational field around the Earth.

The weight of an object depends on the strength of the gravitational field. Different planets have different gravitational field strengths near their surfaces.

The weight of an object is calculated using the equation:

weight = mass × gravitational field strength

$$W = mg$$

> weight, W, in newtons, N
>
> mass, m, in kilograms, kg
>
> gravitational field strength, g, in newtons per kilogram, N/kg

Resultant forces

When two or more forces act on an object, those forces may be replaced by a single force that has the same effect as all the original forces acting together.

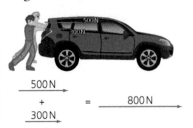

500 N
+
300 N   =   800 N

Figure 5.2 When two forces act in the same direction, they add up to make a larger force. This is called the resultant force.

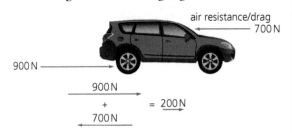

air resistance/drag
700 N

900 N

900 N
+   = 200 N
700 N

Figure 5.3 When two forces act in different directions, a smaller resultant force is produced.

ⓗ Freebody diagrams

A freebody diagram shows all the forces acting on a body. In Figure 5.4 a man, pulling on a rope, has four forces acting on him:

- weight (W)   800 N down
- normal reaction force (R)   800 N up
- the tension from a rope (I) 150 N to the right
- friction (F)   50 N to the left

The resultant force on the man is: 100 N to the right.

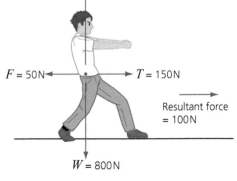

R = 800 N

F = 50 N ← → T = 150 N

Resultant force = 100 N

W = 800 N

Figure 5.4

Resolving forces

A single force can be resolved into two components acting at right angles to each other.

A force of 10 N acts on a box as shown in Figure 5.5. This can be resolved into a vertical component of 6 N and a horizontal component of 8 N.

(a)
10 N
37°

(b)
10 N   6 N
37°
8 N

Figure 5.5

Example

Figure 5.6 shows the direction and size of two forces exerted by two tugboats on a large ship. These are the forces of (i) 50 000 N along OA and (ii) 40 000 N along OB.

Determine by scale drawing the resultant of these two forces.

Answer

Complete the 'parallelogram' of forces by marking in AC (parallel to OA). The resultant force is the red line OC. Using the scale 1 cm = 10 000 N, the resultant force has a length of 7.5 cm on the diagram or 75 000 N.

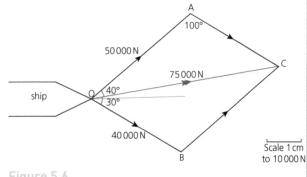

A
100°
50 000 N
75 000 N
C
ship
O
40°
30°
40 000 N
B
Scale 1 cm to 10 000 N

Figure 5.6

Now test yourself

4 Name three contact forces, and three non-contact forces.
5 State the units of:
  (a) weight
  (b) gravitational field strength.
6 An astronaut has a mass of 130 kg in his spacesuit. Calculate his weight on the Moon where the gravitational field strength is 1.6 N/kg.
7 Calculate the resultant forces on the two boxes in Figure 5.7.

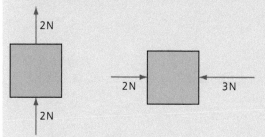

Figure 5.7

8 (a) In Figure 5.8(a), resolve the force of 35 N into a horizontal and a vertical component.

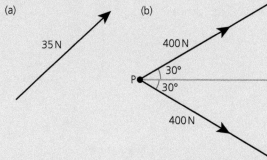

Figure 5.8

(b) In Figure 5.8(b), two forces of 400 N act on point P. Determine the resultant of these two forces by a scale drawing.

Answers on pp. 146–147

Work done and energy transfer

When a force causes an object to move in the direction of the applied force, work is done. For example, when you drag a box along the floor, you do work against a frictional force.

work done = force × distance

$$W = Fs$$

One joule of work is done when a force of one newton causes a displacement of one metre.

1 joule = 1 newton-metre

work done, W, in joules, J

force, F, in newtons, N

distance, s, in metres, m

Energy transfer

When work is done on an object, energy is transferred from one store to another.

For example: when you lift a weight, energy is transferred from the chemical store in your arm to the gravitational potential energy store of the weight (and also to the thermal store in your arm).

Typical mistake

When you hold a weight, without moving it, you get tired; you are transferring energy from your chemical store to a thermal store. But you are not doing any work – you only do work when you move the weight.

Now test yourself

9   State the unit of work.
10 You hold a 20 N weight, without moving it, at arm's length for 5 minutes. Your arm gets tired.
    (a) Discuss what energy transfer takes place.
    (b) State how much work you do while holding the weight.
    (c) Calculate the work done when you lift the 20 N weight through a vertical height of 1.3 m.
11 (a) Calculate the work done when a force of 30 N is used to drag a bag 2.5 m along the floor.
    (b) Discuss the energy transfers in this process.

Answers on p. 147

Forces and elasticity

REVISED

To bend, stretch or change the shape of an object you have to apply at least two forces. In Figure 5.9 a spring is stretched when a force is applied to each end.

If you only apply one force to a spring, you set it in motion but you do not stretch it.

The extension of a spring is directly proportional to the force applied, provided the limit of proportionality is not exceeded (point P on the graph).

force = spring constant × extension

$$F = ke$$

force, F, in newtons, N

spring constant, k, in newtons per metre, N/m

extension, e, in metres, m

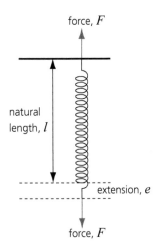

Figure 5.9

This relationship also applies to the compression of an elastic object, where e would be the compression of the object.

Required practical 6 investigates the relationship between force and extension of a spring. You can revisit this by answering Exam practice question 7 on page 72.

Elastic and inelastic deformation

When a spring is stretched elastically, it returns to its original length and shape.

When a spring is stretched beyond the limit of proportionality, the spring is stretched inelastically. This means the spring does not return to its original length and shape.

Energy transfers

A force that stretches a spring does work and elastic potential energy is stored in the spring. Provided the spring is not deformed beyond the limit of proportionality, the work done in stretching the spring is equal to the elastic potential energy stored in the spring.

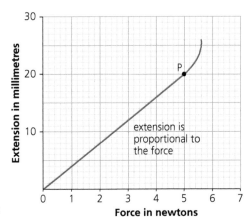

Figure 5.10

When a stretched spring is released, the stored elastic potential energy can be transferred to other energy stores, such as a kinetic store.

The elastic potential energy stored in a spring can be calculated as follows:

elastic potential energy = 0.5 × spring constant × extension²

$$E_e = \frac{1}{2}ke^2$$

elastic potential energy, E_e, in joules, J

spring constant, k, in newtons per metre, N/m

extension, e, in metres, m

12 Explain the terms:
   (a) *elastic deformation*
   (b) *inelastic deformation.*
13 The graph in Figure 5.10 shows the extension of a spring for various applied forces.
   (a) Use the graph to determine how far the spring can be stretched before passing the limit of proportionality.
   (b) Calculate the spring constant in N/m.
   (c) Calculate the work done to stretch the spring by 20 mm.

Answers on p. 147

Moments, levers and gears

REVISED

Moments

When two forces on an object act along the same line, they can squash or stretch the object. But when two forces act along different lines, those forces can rotate the object.

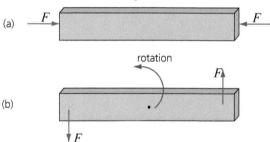

> **Typical mistake**
>
> The unit of a moment is Nm **not** N/m.

Figure 5.11 **(a) Two forces squash an object. (b) Two forces rotate an object.**

The turning effect of a force is called the moment of the force. The size of the moment is defined by the equation:

moment of a force = force × perpendicular distance

$$M = Fd$$

> moment of a force, M, in newton-metres, Nm
>
> force, F, in newtons, N
>
> distance, d, is the perpendicular distance from the pivot to the line of action of the force, in metres, m

Perpendicular means 'at right angles'. Figure 5.12 shows why this distance is important. In Figure 5.12 (a) you get no turning effect when the force acts through the pivot. In Figure 5.12 (b) there is a turning moment that we can calculate:

$$M = Fd$$
$$= 100 \times 0.3$$
$$= 30 \, \text{Nm}$$

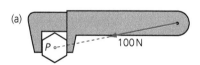

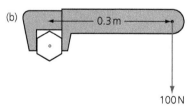

Balancing

If an object is balanced, the total clockwise moment about a pivot equals the total anticlockwise moment about the pivot.

Figure 5.12

In Figure 5.13 two men are lowering a crate into a boat, from a beam. A counterbalance is placed on the other side of the beam to balance the weight of the crate.

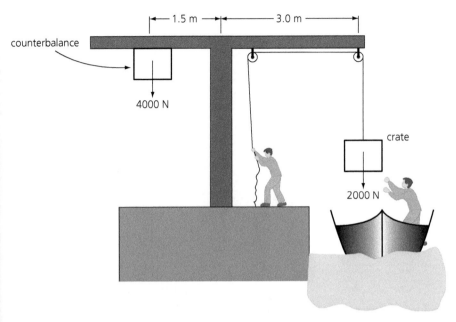

Figure 5.13 Turning moments balance on a beam.

The clockwise moment of the crate is:

$M = 2000 \times 3.0$

$\quad = 6000\,\text{Nm}$

The anticlockwise moment of the counterbalance is:

$M = 4000 \times 1.5$

$\quad = 6000\,\text{Nm}$

You need to be able to calculate the size of a force, or its distance from the pivot, acting on a balanced object.`

Example

Figure 5.14 shows a man using a machine to lift water out of a lake. Calculate the force F.

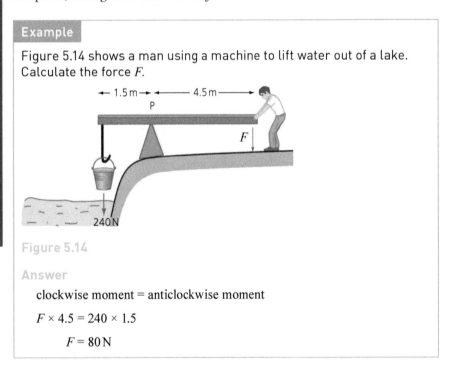

Figure 5.14

Answer

clockwise moment = anticlockwise moment

$F \times 4.5 = 240 \times 1.5$

$\quad\quad F = 80\,\text{N}$

Levers and gears

We use the idea of moments to make life easier for us. Figure 5.15 shows how a lever is used to lift a rock.

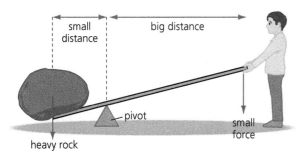

Figure 5.15

The lever transmits the moment applied by the man on the right-hand side to the rock on the left-hand side. So the small force applied by the man with the long lever applies a large force to the rock that is a small distance from the pivot. The beam in Figure 5.14 is also an example of a lever.

Many machines use gears, which can be used to increase or decrease the rotational effects of a force. Figure 5.16 shows how a turning effect can be increased.

The two gears exert equal and opposite forces on each other, but the radius of the output shaft, $2r$, is twice that of the input shaft, r, because it has twice as many teeth.

moment from input shaft $= F \times r = Fr$

moment from output shaft $= F \times 2r = 2Fr$

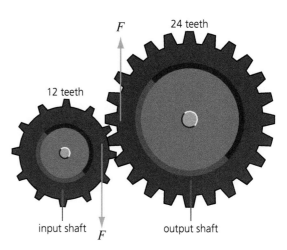

Figure 5.16

Now test yourself

TESTED

14 The correct unit for a moment is:
  A N/m   B N/m²   C Nm   D m/N

15 Calculate the moment due to the weight of the flower basket about P.

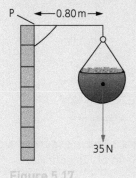

Figure 5.17

16 Explain why you use a long lever to lift a rock.

17 Calculate the force that the hammer exerts on the nail to pull it out.

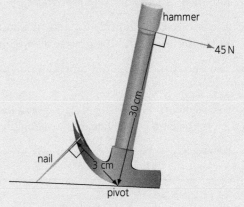

Figure 5.18

18 Figure 5.19 shows a boy and girl who are balanced on a seesaw. Calculate the weight of the boy.

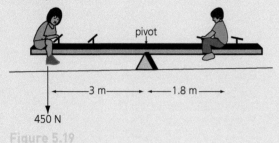

pivot

—3 m— —1.8 m—

450 N

Figure 5.19

19 Figure 5.20 shows an 'axle and wheel', a device which can be used to lift a heavy load.
(a) Calculate the moment of the 750 N load about the axle AB.
(b) Use the idea of moments to calculate the force F required to lift the load.

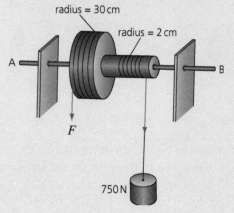

radius = 30 cm

radius = 2 cm

A

B

F

750 N

Figure 5.20

Answers on p. 147

Pressure and pressure differences in fluids

A fluid is a gas or a liquid.

Figure 5.21 shows gas in a cylinder. The pressure in a fluid causes a force normal (at right angles) to any surface.

The pressure can be calculated using the equation:

$$\text{pressure} = \frac{\text{force normal to a surface}}{\text{area of that surface}}$$

$$P = \frac{F}{A}$$

pressure, P, in pascals, Pa

force, F, in newtons, N

area, A, in metres squared, m^2

⊕ Pressure depends on depth and density

The pressure gets larger as you dive down to the bottom of a swimming pool, because there is a greater weight of water on top of you.

The pressure exerted by a liquid also depends on the density of that liquid – the denser the liquid the greater its weight.

The pressure due to a column of liquid can be calculated using the equation:

$$\text{pressure} = \text{height of} \times \text{density of} \times \text{gravitational field}$$
$$\text{the column} \quad \text{the liquid} \quad \text{strength}$$

$$P = h\rho g$$

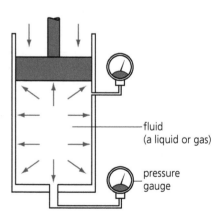

fluid (a liquid or gas)

pressure gauge

Figure 5.21 The pressure in a fluid acts in all directions.

pressure, P, in pascals, Pa

height, h, in metres, m

density, ρ, in kilograms per metre cubed, kg/m^3

gravitational field strength, g, in newtons per kilogram, N/kg

H

Example

Calculate the pressure experienced by a diver at a depth of 25 m in the sea. The density of the seawater is 1020 kg/m³, atmospheric pressure is 101 000 Pa, the gravitational field strength is 9.8 N/kg.

Answer

$P = h\rho g$

$\quad = 25 \times 1020 \times 9.8$

$\quad = 249\,900\,\text{Pa}$

So the total pressure including atmospheric pressure is:

$249\,900\,\text{Pa} + 101\,000\,\text{Pa}$

$= 250\,900\,\text{Pa}$

$= 250\,\text{kPa to 2 significant figures}$

Exam tip

Note that the final answer is given to 2 significant figures, to match the data given.

Floating

A partially (or totally) submerged object experiences a greater force on the bottom surface than on the top surface.

The resultant of the two forces on the cylinder is called the upthrust, U.

- An object floats when the upthrust from the fluid is equal to the object's weight. When the weight of the object is greater than the upthrust, the object sinks.
- An object sinks when its density is greater than that of the fluid.
- An object floats when its density is equal or less than the density of the fluid.

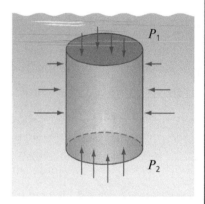

Figure 5.22 The cylinder experiences an upward force from the water because P_2 is greater than P_1.

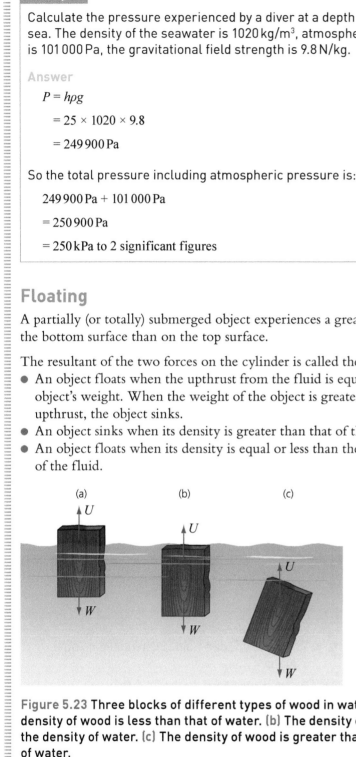

Figure 5.23 Three blocks of different types of wood in water. (a) The density of wood is less than that of water. (b) The density of wood equals the density of water. (c) The density of wood is greater than the density of water.

Atmospheric pressure

The Earth's atmosphere is confined to a thin layer above the Earth's surface.

The atmosphere gets less dense with increasing altitude.

Air molecules exert a pressure by colliding with a surface.

At high altitudes there are fewer molecules (in a given volume). So the atmospheric pressure decreases with an increase in height.

Now test yourself

20 State the unit of pressure.
21 (a) Explain how the atmosphere exerts a pressure.
    (b) Explain why atmospheric pressure is reduced at a high attitude.
22 (a) Explain why the pressure increases on a diver as he goes deeper in the sea.
    (b) The pressure at the bottom of a column of mercury, 76.0 cm high, is 101 300 Pa. The gravitational field strength is 9.8 N/kg. Calculate the density of mercury.
23 A piece of wood has a weight of 16 N. When lowered into a bowl of water the wood experiences an upthrust of 14 N.
    (a) Draw a diagram to show the forces acting on the wood.
    (b) (i) Explain whether the wood will float or sink.
         (ii) Is the density of wood greater or less than the density of water?

Answers on p. 147

Summary

- Scalar quantities have magnitude only.
- Vector quantities have magnitude and directions.
- A force is a push or a pull; there are contact and non-contact forces.
  Force is measured in newtons, N.
- Weight is the pull of gravity on a mass.

  weight = mass × gravitational field strength

  $$W = mg$$

- A resultant force is the vector sum of a number of forces acting on a body.
- The work done by a force acting on an object is calculated by:

  work done = force × distance (moved along the line of action of the force)

  $$W = Fs$$

  Work done is measured in joules, J.
- The extension of an elastic object such as a spring is proportional to the force, provided the spring does not exceed its limit of proportionality.

  force = spring constant × extension

  $$F = ke$$

  k is measured in N/m.

- The work done in stretching (or compressing) a spring (up to the limit of proportionality) is calculated using:

  elastic potential energy $= \dfrac{1}{2} \times$ spring constant × (extension)^2

  $$E_e = \frac{1}{2}ke^2$$

- Moment has the unit Nm:

  moment of force = force × perpendicular distance

  When an object is balanced, the total clockwise moment about a pivot equals the total anticlockwise moment.
- Fluids exert pressures at right angles to the surface of their container.

  pressure $= \dfrac{\text{force normal to a surface}}{\text{area of the surface}}$

  $$P = \frac{F}{A}$$

- Pressure under a fluid is calculated using the equation:

  pressure = height × density × gravitational field strength

  $$P = h\rho g$$

Exam practice

1 Which one of the following is a vector quantity? [1]
   force   speed   energy

2 How many newtons are there in a kilonewton? [1]
   10   1000   1000000

3 Which of the following is the correct unit for energy? [1]
   newton   joule   watt

4 An astronaut in his spacesuit has a mass of 120 kg on the Earth, where the gravitational
   field strength is 9.8 N/kg.
   (a) Calculate his weight. [2]
   (b) He climbs up a ladder 4.5 m high into the spacecraft.
      Calculate the work done by his legs as he climbs. [3]
   (c) Explain why the astronaut has a smaller weight when he lands on Mars. [2]

5 A student sets up the apparatus below to investigate the principle of moments.
   She hangs a 4 N weight from the 50 cm mark on the ruler.
   She uses a force meter to hold the ruler in a horizontal position.
   The force meter reads from 0 N to 20 N.

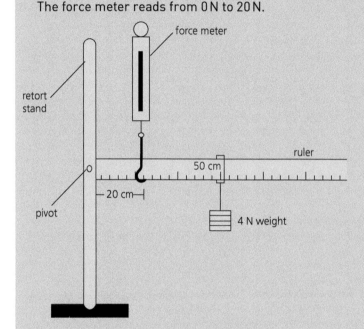

Figure 5.24

   (a) Explain how the student checks that the ruler is horizontal. [2]
   (b) Calculate the moment of the 4 N weight about the pivot. [3]
   (c) The student holds the ruler horizontal with the force meter on the 20 cm mark.
      She calculates that the meter will read 10 N.
      (i) Show how she reached the answer of 10 N. [2]
      (ii) The actual reading is 12.5 N. Explain why the correct reading should be higher than 10 N. [2]

6 A diver is swimming at a depth of 15 m below the surface of the sea.
   The atmospheric pressure is 100 000 Pa.
   (a) Calculate the total pressure on the diver. The density of the seawater is 1020 kg/m³ and the
      gravitational field strength is 9.8 N/kg. [4]
   (b) At another depth the pressure on the diver's facemask is 200 000 Pa. The area of his face
      mask is 0.018 m². Calculate the force exerted by the water on the outside of his mask. [2]

Required practical 6

7 A student carried out an investigation into the stretching of a spring A. He makes a hypothesis that the extension of the spring will be proportional to the weight on it.

Figure 5.25 shows the spring in (i) its unstretched state and (ii) when it has been stretched by a load.

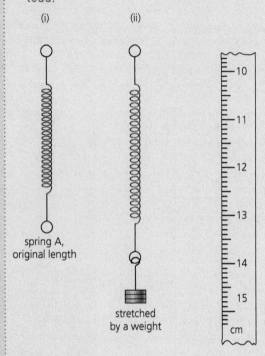

(i)

(ii)

spring A,
original length

stretched
by a weight

Figure 5.25

(a) Use Figure 5.25 to calculate the spring's extension in this case. [2]

(b) The student uses weights up to 5 N to stretch spring A. The results are shown in Figure 5.26.

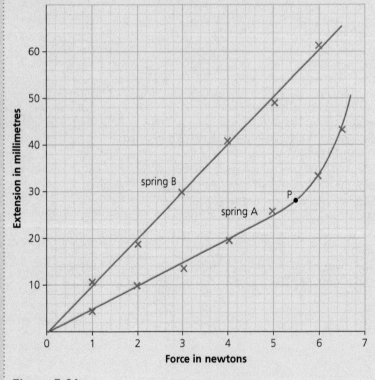

Figure 5.26

(i) Explain whether this graph supports the student's hypothesis. [2]

(ii) Not all the points plotted by the student lie on the straight line. Explain what type of error caused this and how you would attempt to reduce this type of error. [2]

(iii) The student increases the load to 6.5 N. Explain what happens to the spring now. [1]

(iv) Calculate the spring constant for spring A. Express your answer in N/m. [3]

(c) The student repeats his experiment for a second spring B (Figure 5.27). His results are also shown in Figure 5.26.

(i) State which spring is stiffer. Explain your answer. [2]

(ii) Use the graph to predict the extension of both springs together when a load of 4 N is used to extend them.

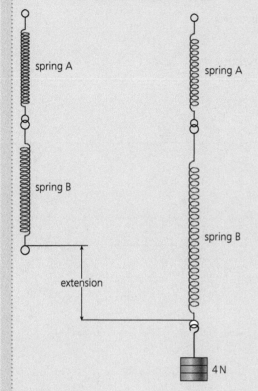

Figure 5.27

(d) Hooke's law states that 'the extension of a spring is proportional to the applied force'. This is an example of a law named after a scientist. Explain why Hooke's law is now accepted to be true for all springs (over a limited range of applied forces). [2]

Answers and quick quiz 5A online

ONLINE

5B Forces and motion

Describing motion along a line

Distance and displacement

Distance is how far an object moves. Distance does not involve direction. Distance is a scalar quantity.

Displacement includes both the distance an object moves, measured in a straight line from the start point to the finish point and the direction of that straight line.

Displacement is a vector quantity.

> **Example**
>
> Figure 5.28 shows the path taken by a walker. She walks 8 km due east, (AB), then 6 km due south (BC).
>
> The distance walked is 8 km + 6 km = 14 km.
>
> The displacement is shown by the vector AC. This is 10 km along a direction 37° south of east.
>
>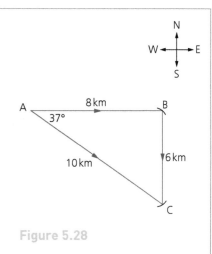
>
> Figure 5.28

Speed and velocity

For an object moving at a constant speed, the distance travelled in a specific time can be calculated using the equation:

distance travelled = speed × time

$$s = vt$$

As speed does not involve direction, it is a scalar quantity.

Velocity is the speed of an object in a given direction. Velocity is a vector quantity.

> distance, s, in metres, m
>
> speed, v, in metres per second, m/s
>
> time, t, in seconds, s

Typical speeds

The speed at which a person can walk, run or cycle depends on their age, fitness, how far they have already travelled and the nature of the ground being travelled over.

Typical values may be taken as:
- walking   1.5 m/s
- running   3 m/s
- cycling   6 m/s

H Motion in a circle

Figure 5.29 shows a satellite in a circular orbit around the Earth. The satellite moves at a constant speed, but its velocity changes all the time. The velocity changes because the direction of travel changes. The pull of the Earth's gravity changes the satellite's direction of travel.

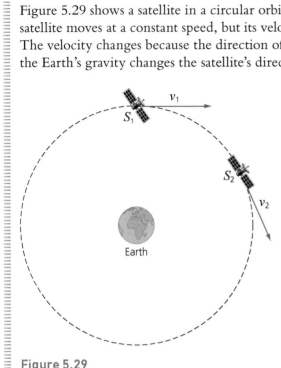

Figure 5.29

Distance–time graphs

When an object moves along a straight line, we can represent how far it has travelled by a distance–time graph.

Figure 5.30 shows a distance–time graph for a runner.

The speed of the runner is calculated using the gradient of the graph.

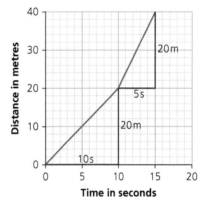

Figure 5.30

Example

Using the graph in Figure 5.30, calculate the speed of the runner between 10 and 15 seconds.

Answer

The speed from 10 s to 15 s is:

$$\text{speed} = \frac{\text{distance}}{\text{time}}$$

$$= \frac{20}{5}$$

$$= 4 \, \text{m/s}$$

When an object is accelerating, the gradient of the distance–time graph changes.

We calculate the gradient by drawing a tangent to the curve.

In Figure 5.31, the gradient at A is:

$$\text{speed} = \frac{\text{distance}}{\text{time}}$$

$$= \frac{75}{25}$$

$$= 3\,\text{m/s}$$

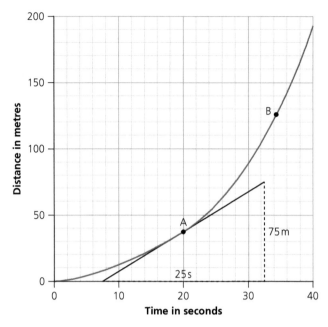

Figure 5.31

Now test yourself

1 An athlete runs (i) 100 m in 10.0 s and (ii) 400 m in 46.0 s.
   (a) Calculate his speed in each case.
   (b) Explain why one speed is faster than the other.
2 Figure 5.32 shows a distance–time graph for a car travelling along a straight road.

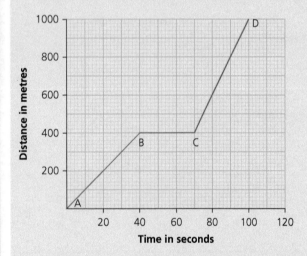

Figure 5.32

   (a) How far did the car travel over the region CD?
   (b) Over which part of the journey did the car travel at its greatest speed? Explain your answer.
   (c) For how long was the car stopped?
   (d) Calculate the car's speed over the region AB.
3 Calculate the speed of the moving object at point B in the distance–time graph in Figure 5.31.
4 A train travels at a constant speed of 55 m/s.
   (a) Calculate the distance travelled by the train in 1 minute.
   (b) Calculate the time it takes the train to travel 11 km.

Answers on pp. 147–148

Acceleration

The average acceleration of an object is calculated using the equation:

$$acceleration = \frac{change\ in\ velocity}{time\ taken}$$

$$a = \frac{\Delta v}{t}$$

When an object slows down it is decelerating.

> acceleration, a, in metres per second squared, m/s^2
>
> change in velocity, Δv, in metres per second, m/s
>
> time, t, in seconds, s

Velocity–time graphs

Figure 5.33 shows a velocity–time graph for a cyclist.
- He accelerates up to 12 m/s over the first 8 seconds – section AB.
- He travels at a constant speed for the next 12 seconds – section BC.
- Then he decelerates from 12 m/s over the last 4 seconds – section CD.

> **Typical mistake**
>
> Often students give the unit of acceleration as m/s – it is **m/s²**. An acceleration is a change in velocity, m/s, in a given time, s. So the unit is m/s divided by s – **m/s²**.

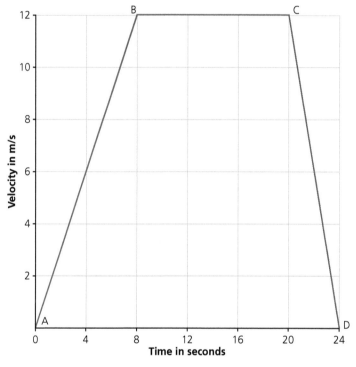

Figure 5.33

You can calculate the acceleration from a velocity–time graph.

Acceleration over time AB:

$$a = \frac{\Delta v}{t}$$

$$= \frac{12}{8}$$

$$= 1.5\ m/s^2$$

You can work out the distance travelled from a velocity–time graph.

5B Forces and motion

Examples

Calculate how far the cyclist travelled:

1   when he travelled at a constant speed
2   while he accelerated.

Answers

1   $d = vt$

      $= 12 \times 12$

      $= 144\,m$

We can also work out the distance by counting the squares: under the line BC there are $6 \times 3 = 18$ squares.
But each square has a value $= 2\,m/s \times 4\,s = 8\,m$.
So the distance travelled $= 18 \times 8 = 144\,m$

2   $d =$ average speed × time

      $= 6 \times 8$

      $= 48\,m$

Note, here we have to use the average speed which is 6 m/s – half-way between 0 and 12 m/s.
Or we could have worked out the distance using the 'area' under the graph:

'area' $= \dfrac{1}{2} \times 12\,m/s \times 8\,s = 48\,m$

This is equivalent to 6 squares.
In a Higher Tier exam question, you might be asked to work out the distance travelled for a graph with a curved line. Question 8(c) on page 79 is an example of this.

Exam tip

The gradient of a velocity–time graph equals the acceleration.

Exam tip

The area under a velocity–time graph equals the distance travelled.

Equation of motion for uniform acceleration

The following equation applies to uniform acceleration:

   (final velocity)² − (initial velocity)² = 2 × acceleration × distance

$$v^2 - u^2 = 2as$$

final velocity, v, in metres per second, m/s

initial velocity, u, in metres per second, m/s

acceleration, a, in metres per second squared, m/s²

distance, s, in metres, m

Example

A light aircraft accelerates from rest along a runway. The aircraft takes off at a speed of 35 m/s having travelled 450 m along the runway. Calculate the aircraft's acceleration.

Answer

   $v^2 - u^2 = 2as$

  $35^2 - 0 = 2 \times a \times 450$

      $a = \dfrac{35^2}{900}$

         $= \dfrac{1225}{900}$

         $= 1.4\,m/s^2$

Terminal velocity

When an object falls freely under gravity, near the Earth's surface, it has an acceleration of $9.8 \, m/s^2$.

When an object falls through a **fluid**, resistive forces act on the object. Initially the acceleration due to gravity is $9.8 \, m/s^2$. But as the object increases its speed, larger resistive forces act on it; so the resultant force on the object decreases and the acceleration decreases. Eventually the resultant force is zero and the object moves at a constant velocity. This is called the **terminal velocity**.

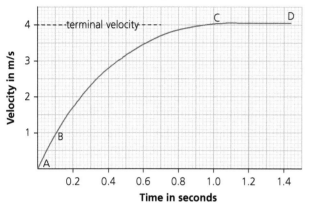

Figure 5.34 Velocity–time graph for a ball bearing falling through a fluid.

Figure 5.34 shows a velocity–time graph for a ball bearing falling through a fluid.

● Over the region AB it accelerates under gravity at $9.8 \, m/s^2$.
● Over the region BC the acceleration decreases as drag forces increases.
● Over the region CD the ball bearing falls at its terminal velocity. There is no resultant force and no acceleration.

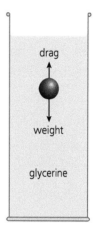

Figure 5.35 A ball bearing falls through glycerine at a constant (terminal) velocity. The drag force is balanced by the weight.

Now test yourself

TESTED ☐

5 State the correct units for
   (a) velocity
   (b) acceleration.
6 (a) A runner accelerates from a speed of 3 m/s to 5 m/s in a time of 8 seconds. Calculate her acceleration.
   (b) A car slows down from a speed of 30 m/s to a speed of 18 m/s in a time of 3 seconds. Calculate the deceleration of the car.
7 This question refers to the velocity–time graph in Figure 5.33.
   (a) Calculate the deceleration of the cyclist in the last 4 seconds of his journey.
   (b) Calculate the distance travelled while the cyclist decelerates.
8 Figure 5.34 shows the velocity–time graph for a ball bearing falling through a fluid.
   (a) Calculate the ball bearing's average acceleration over the times:
       (i) 0 to 0.1 s
       (ii) 0.3 to 0.4 s
   (b) Explain why the ball bearing reaches a terminal velocity.
   (c) The ball bearing reaches the bottom of the container after 1.4 s. Which of the following is closest to the length of the container? Explain your answer.
       1.5 m   4.5 m   7.5 m
9 A train slows down from a speed of 50 m/s to 30 m/s over a distance of 1 km. Calculate the train's deceleration.

Answers on p. 148

Forces, accelerations and Newton's laws of motion

Newton's first law

Newton's first law states that:

If the resultant force acting on an object is zero:
● and the object is stationary, the object remains stationary
● and when the object is moving, the object continues to move with the same speed in a straight line.

Newton's first law tells us that the velocity of an object only changes if a resultant force acts on an object. When a resultant force acts, a moving body can:
● speed up
● slow down
● or change direction.

When a car travels at a constant speed along a road, the resistive forces on the car balance the driving force on the car.

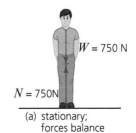

(a) stationary; forces balance

(b) moving at steady speed; forces balance

Figure 5.36

Inertia

The word *inertia* comes from the Latin word meaning *inactivity* or *inaction*.

In physics, we use the word *inertia* to describe an object's tendency to remain at rest or to continue moving at a constant speed.
Inertial mass can be defined as the ratio of force over acceleration:

$$m = \frac{F}{a}$$

$$\text{Inertial mass} = \frac{\text{force}}{\text{acceleration}}$$

Newton's second law

The acceleration of an object is proportional to the resultant force acting on the object and inversely proportional to the mass of the object.

The statement above can be expressed mathematically in this form:

$$a \propto F$$

$$a \propto \frac{1}{m}$$

Newton's second law may also be written as an equation:

resultant force = mass × acceleration

$$F = ma$$

Typical mistake

When using Newton's second law make sure you have calculated the **resultant** force.

force, F, in newtons, N

mass, m, in kilograms, kg

acceleration, a, in metres per second squared, m/s²

Example

800 N   500 N

Figure 5.37

The mass of the car is 1000 kg; calculate its acceleration.

Answer

$$F = ma$$

$$800 - 500 = 1000 \times a$$

$$a = \frac{300}{1000}$$

$$= 0.3 \,\text{m/s}^2$$

Typical masses, accelerations and forces

You should know the approximate sizes of speeds, accelerations and forces involved in everyday transport. The table gives some examples for a car and a train.

	Mass	Acceleration from rest	Resultant force acting to accelerate from rest	Maximum speed
car	≈ 1500 kg	≈ 2 m/s²	≈ 3 000 N	≈ 30 m/s
train	≈ 200 × 10³ kg	≈ 0.2 m/s²	≈ 40 000 N	≈ 55 m/s

Maths note

You should be familiar with these mathematical symbols:
- proportional to: α
- approximately equal to: ≈

Required practical 7

The effect of mass and force on acceleration

Figure 5.38 shows the apparatus you might use to investigate how acceleration of an object depends on the object's mass and the force applied to accelerate it.

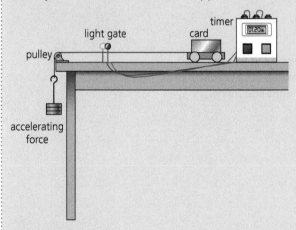

Figure 5.38

Measuring acceleration

The trolley is allowed to accelerate from rest.

The trolley's final speed is calculated:

$$v = \frac{\text{length of card}}{\text{time taken to pass through light gate}}$$

The trolley's acceleration is calculated:

$$a = \frac{\text{final speed, } v}{\text{time taken to reach the light gate}}$$

Changing the force

The mass of the trolley is kept constant.

The force is varied and accelerations are calculated for different forces to show that:

$a \propto F$

Doubling the force, for a fixed mass, doubles the acceleration.

Changing the mass

The force accelerating the trolley is kept constant.

The mass is varied and accelerations calculated for different masses to show that:

$a \propto \dfrac{1}{m}$

Doubling the mass, for a fixed force, halves the acceleration.

Newton's third law of motion

Newton's third law states that whenever two objects interact, the forces they exert on each other are equal and opposite.

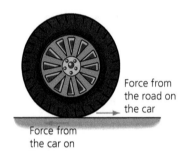

(a) If I push you with a force of 100 N, you push me back with a force of 100 N.

(b) When the wheel of a car turns, it pushes the road backwards. The road pushes the wheel forwards with an equal and opposite force.

(c) A spacecraft orbiting the Earth is pulled downwards by the Earth's gravity. The spacecraft exerts an equal and opposite gravitational force on the Earth.

(d) Two balloons have been charged positively. They each experience a repulsive force from the other. These forces are of the same size, so each balloon (if of the same mass) is lifted through the same angle.

Figure 5.39

These are the features of Newton's third law pairs:
- they act on separate bodies
- they are always of the same type – for example, two gravitational forces or two contact forces
- they are of the same magnitude
- they act along the same line
- they act in opposite directions.

Now test yourself

TESTED ☐

10 A feather falls to the ground at a constant speed. Which of the following is true?
  A The feather's weight is slightly greater than the air resistance on the feather.
  B The feather's weight is equal to the air resistance acting on it.
11 (a) A book is at rest on a table.
  (i) Draw a diagram to show the two forces acting on the book – its weight, W, and the contact force from the table, R.
  (ii) Explain why the resultant force on the book is zero.
  (b) A student says:

  'Equal and opposite forces act on the book; this is an example of Newton's third law.'

  (i) Explain why the student is wrong.
  (ii) What is the equal and opposite force to the weight of the book, as described by Newton's third law?
12 Explain why a racing car is designed to have as low a mass as possible.
13 A car has a mass of 1500 kg. It accelerates from 10 m/s to 18 m/s in 12 s.
  (a) Calculate the car's acceleration.
  (b) Calculate the resultant force acting on the car while it accelerates.
14 A boy blows up a balloon, and then releases it, so that air escapes from it. The balloon flies around the room. Use Newton's third law to explain the motion of the balloon.
15 The diagram shows a speed boat.
  (a) Explain why the boat is moving at a constant speed.

8000 N
forward force

8000 N
drag force

Figure 5.40

  (b) The engine speed is reduced so that the boat slows down. Use the information in the graph in Figure 5.41 to calculate the boat's deceleration over the region AB.

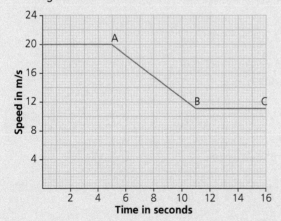

Figure 5.41

  (c) (i) The mass of the boat is 2500 kg. Calculate the resultant force on the boat while it decelerates.
  (ii) Calculate the forwards force on the boat due to the propellers, while the boat decelerates.
  (iii) State the size of the drag force acting on the boat over the region BC.

Answers on p. 148

Driving: stopping distance

REVISED

When a driver of a vehicle sees a hazard she reacts, applies the brakes and stops the vehicle.
- stopping distance = thinking distance + braking distance
- thinking distance = distance travelled while the driver reacts
- braking distance = distance travelled while the car brakes
- stopping distance = total distance travelled from when the driver first sees a hazard to the point where the car stops

Factors affecting reaction times

Reaction times vary from person to person; typical times vary from 0.2 s to 0.9 s.

Driver's reaction times are slower if they:
- have been drinking alcohol
- have been taking certain types of drugs
- are tired
- are distracted by using their mobile phone.

> **Exam tip**
>
> A common exam question is to ask you about the factors that affect reaction times and braking distance.

Factors affecting braking distance

- **Speed**. A large speed increases the braking distance.
- **Force**. A large braking force reduces the braking distance.
- **Mass**. Making the car more massive by carrying a large load increases the braking distance.
- **Weather**. In wet or icy conditions there is less friction between the road and tyres – braking distance increases.
- **Vehicle maintenance**. Worn brakes or worn tyres increase the braking distance.
- **Road condition**. Some road surfaces affect the friction on the tyres – a smooth or muddy road can increase the braking distance.

Braking and kinetic energy

The work done by the brakes reduces a vehicle's kinetic energy, causing the vehicle to stop. The kinetic energy is transferred to the thermal store in the brakes.

work done = kinetic energy transferred

$$Fs = \frac{1}{2}mv^2$$

This shows the braking distance is proportional to v^2 (and the mass of the car). If the speed doubles, the braking distance increases four times.

Now test yourself

TESTED

16 Which of the following affects the thinking distance of a driver?
  an icy road   a drunk driver   worn brakes
17 Which of the following affect the braking distance of a car?
  a tired driver   a muddy road   the speed of the car
18 A car is travelling at 15 m/s. A driver has a reaction time of 0.4 s.
  (a) Calculate the thinking distance of the driver.
  (b) When the driver is tired, his reaction time increases to 0.6 s. Calculate the thinking distance now.

Answers on pp.148–149

Ⓗ Momentum

Momentum is defined by the equation:

momentum = mass × velocity

$$p = mv$$

Since velocity is a vector quantity, momentum is also a vector quantity.

> momentum, p, in kilogram metres per second, $kg\,m/s$
>
> mass, m, in kilograms, kg
>
> velocity, v, in metres per second, m/s

Changes in momentum

$$force = \frac{change\ of\ momentum}{time}$$

$$F = \frac{m\Delta v}{\Delta t}$$

So, force is the rate of change of momentum.

This tells us that when momentum changes occur in a short time, large forces act.

Example

A child jumps off a wall. Explain why he bends his legs on landing.

Answer

$$F = \frac{m\Delta v}{\Delta t}$$

The momentum change of the child is determined only by the height of the wall. By bending his legs, the time, Δt, in which he stops increases. Therefore, the force acting to slow the child decreases, and he is less likely to get hurt.

Exam tip

Note that in the answer given in the example the equation $F = \frac{m\Delta v}{\Delta t}$ was used, even though there was no calculation to perform. This is a good way to show the examiner you know what you are talking about.

Many safety features increase the time in which moving objects come to a stop.

- Crumple zones in cars increase the time in which a car stops in a crash.
- Seat belts in cars make sure the passengers use all of the stopping time in a crash. Without a seat belt, you keep moving until you stop very quickly.
- Shin pads, worn by hockey or football players, stop a ball, stick or foot in a longer time.

Conservation of momentum

In a closed system, the total momentum before an event is equal to the total momentum after the event.

The word 'closed' means that no external forces act on the system. For example, if a moving car collides with a stationary car and they stick together, they share the momentum. Both cars move together, but at a lower speed than the moving car had before the collision.

Example

Figure 5.42 shows a collision taking place between two trolleys in a laboratory. After the collision, the trolleys stick together and move off to the right. Calculate the speed after the collision.

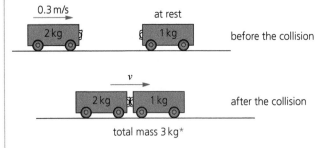

Figure 5.42

Answer

momentum before the collision = 2 × 0.3

= 0.6 kg m/s

momentum after the collision = 0.6 kg m/s

0.6 = 3 × v (total mass 3 kg)*

v = 0.2 m/s

Momentum as a vector quantity

Since momentum is a vector quantity we must take account of the direction of the motion.

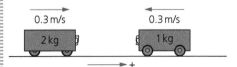

Figure 5.43

Example

Refer to the two trollies in Figure 5.43. Calculate what happens if they approach each other, each travelling at 0.3 m/s. They stick together after the collision.

Answer

First we must define a positive direction; here positive means travelling to the right.

momentum before the collision = +2 × 0.3 − 1 × 0.3

= +0.3 kg m/s

momentum after the collision = 0.3 kg m/s = 3 × v (total mass 3 kg)

v = 0.1 m/s

This means the trolleys move to the right at 0.1 m/s after the collision.

Now test yourself

19 Which of the following is the correct unit for momentum?

kg m s   kg m/s²   kg m/s

20 A car of mass 1250 kg travels at a speed of 20 m/s. Calculate the car's momentum.

21 Use your knowledge of physics to explain the following safety features:
  (a) passengers wear seat belts in cars
  (b) cyclists wear helmets.

22 A cricket player catches a ball that is travelling in a horizontal direction at a speed of 30 m/s. She moves her hands backwards to catch the ball in a time of 0.1 s. The cricket ball has a mass of 160 g.
  (a) Calculate the force exerted by the player's hands on the ball as it stops.
  (b) Explain why the player moves her hands backwards as she catches the ball.

23 A trolley of mass 1.5 kg is travelling to the right with a speed of 0.60 m/s. It collides with a stationary trolley of mass 1.0 kg and sticks to it.
  Calculate the speed of the trolleys after the collision.

Answers on p. 149

Summary

- distance travelled = speed × time
- When the speed changes during the motion, an average speed may be calculated:

$$\text{average speed} = \frac{\text{distance travelled}}{\text{time}}$$

- The gradient of a distance–time graph is equal to the speed of the object travelling.

- $$\text{acceleration} = \frac{\text{change of velocity}}{\text{time}}$$

- The gradient of a velocity–time graph is equal to the acceleration.
- The 'area' under a velocity–time graph is equal to the distance travelled.
- The following equation applies to uniform motion:

$$(\text{final velocity})^2 - (\text{initial velocity})^2 = 2 \times \text{acceleration} \times \text{distance}$$

$$v^2 - u^2 = 2as$$

- Acceleration due to Earth's gravity: $g = 9.8 \text{ m/s}^2$.
- A falling object reaches a terminal velocity when the resistive forces balance the object's weight.
- Newton's first law states that when the resultant force on an object is zero: the object remains at rest or moves with a constant speed in a straight line.

- Newton's second law states that:

resultant force = mass × acceleration

$$F = ma$$

- Newton's third law states that: whenever two objects interact, the forces they exert on each other are equal and opposite.

- stopping distance = thinking distance + braking distance

- Reaction times are slowed if a driver is tired, consumes alcohol or some drugs, or is distracted.
- Braking distances are larger at high speeds, when carrying a massive load, in wet or icy conditions, or if the tyres or brakes of a vehicle are worn.

- momentum = mass × velocity

- You should be able to use this equation to explain safety features such as crash helmets and seat belts:

$$\text{force} = \frac{\text{change of momentum}}{\text{time}}$$

- In a closed system, the total momentum of the system is conserved.

Exam practice

1 Figure 5.44 shows a distance–time graph for part of a car's journey.

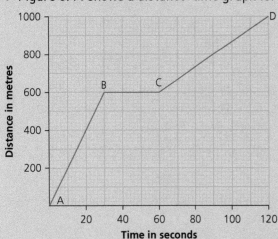

Figure 5.44

(a) How long was the car stopped for? [1]
(b) Over which part of the journey was the car travelling fastest? Explain your answer. [2]
(c) Calculate the average speed for the whole journey from A to D. [2]

2 A small ball with a weight of 0.5 N is allowed to fall from rest. It accelerates until it reaches a constant velocity.

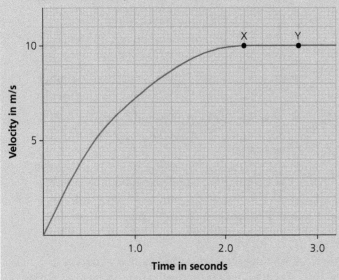

Figure 5.45

(a) Use the graph to determine the time it took the ball to reach a constant velocity. [1]
(b) Explain the shape of the graph. [3]
(c) Calculate the acceleration of the ball in the first 0.4 seconds that it falls. [3]
(d) Calculate the distance the ball travels between the times marked X and Y on the graph. [2]
(e) State the size of the resultant force acting on the ball when it travels at a constant velocity. Explain your answer. [2]

3 A car driver is travelling at 25 m/s on a motorway when she sees a hazard ahead. The car travels a further 15 m, before the driver presses the brake pedal.
(a) Calculate the driver's reaction time. [2]
(b) State two factors that could affect her reaction time. [2]
(c) The car comes to a halt after a distance of 60 m. Use the equation

$$v^2 - u^2 = 2as$$

to calculate the car's deceleration. [3]
(d) State two factors which could affect the car's braking distance. [2]

(e) On another occasion the car is travelling at 12.5 m/s. State whether the braking distance will be more than, equal to, or less than 30 m. Explain your answer. [2]

4 A skydiver has a mass of 100 kg including his parachute. Gravitational field strength g = 9.8 N/kg.
  (a) Calculate the skydiver's weight. [2]
  (b) Before he opens his parachute he falls with a constant velocity of 50 m/s. State what the resultant force acting on him is. [1]
  (c) The skydiver opens his parachute, which causes the drag force on him to increase to 1480 N.
    (i) State the resultant force acting on a skydiver now. [1]
    (ii) Calculate the deceleration of the skydiver when he opens the parachute. [2]

5 Use Newton's laws of motion to explain each of the following:
  (a) You can throw a tennis ball of mass 60 g much further than you can put a shot of mass 4 kg. [3]
  (b) Two ice skaters, Paul and Jane, stand next to each other on the ice. Paul gives Jane a push. Both ice skaters begin to move. Explain why. [3]
  (c) A parcel is placed on the seat of a car. When the car brakes suddenly the parcel falls onto the floor. Explain why. [3]

6 Figure 5.46 (a) shows a rocket and a Mars lander, which are at rest in deep space. The mass of the rocket is 80 000 kg and the mass of the Mars lander is 16 000 kg. Rockets are ignited to separate the rocket and the lander.
  (a) Figure 5.46 (b) shows the rocket and lander after separation. Calculate the momentum of the lander. [3]
  (b) State the momentum of the rocket after separation. [1]
  (c) Calculate the velocity of the rocket. [2]

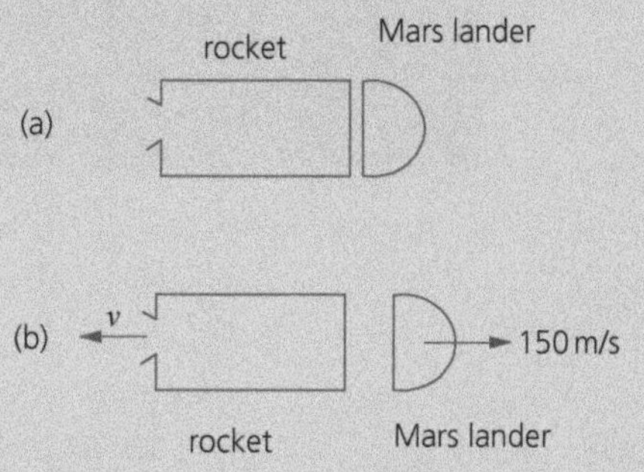

Figure 5.46

7 Describe an experiment to measure the acceleration of an object in a laboratory. Describe the apparatus you would use and explain what measurements you would take to calculate the acceleration. Explain what action you would take to ensure the experiment is safe and does not damage any apparatus. [6]

Answers and quick quiz 5B online

ONLINE

6 Waves

All waves carry energy and information. Sound waves allow us to hear, and electromagnetic waves let us see and enable communications via phone, internet, television and radio.

We study waves on water and on ropes or springs, so that we can apply that understanding to sound and electromagnetic waves, which we cannot see.

Waves in air, fluids and solids

REVISED

When a stone lands in a pond, you see water ripples spreading out. The circular ripples give us the information about where the stone landed (if you did not see it). The energy is transmitted through the water, but the water itself does not move outwards. These ripples are examples of waves.

Transverse waves

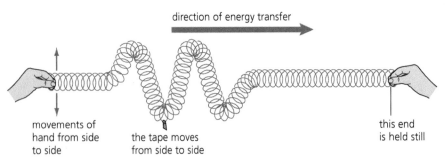

Figure 6.1

Figure 6.2 **The coloured tape shows that the slinky coils move from side to side, like the hand.**

Figure 6.2 shows a **transverse wave**. The student moves the slinky from side to side, and the energy is transmitted along the spring. The water waves in Figure 6.1 are also examples of transverse waves. Electromagnetic waves (including light) are transverse waves.

> A **transverse wave** is one in which the vibrations causing a wave are at right angles to the direction of energy transfer.

Longitudinal waves

Figure 6.3 shows a **longitudinal wave**. Energy is transmitted along the slinky when it is pushed backwards and forwards. In some places coils are pushed together – these are compressions (C); in other places the coils are pulled apart, these are rarefactions (R). Sound waves are examples of longitudinal waves.

> A **longitudinal wave** is one in which the vibration causing the wave is parallel to the direction of energy transfer.

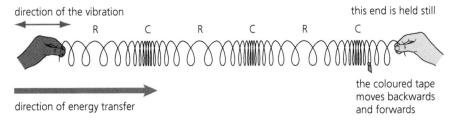

Figure 6.3 **The coloured tape shows that the slinky coils move backwards and forwards like the hand.**

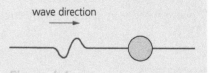
Properties of waves

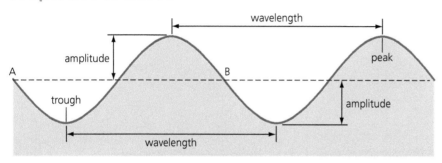

Figure 6.5

- The **amplitude**, A, is the distance from a wave peak to the middle, or from a wave trough to the middle.
- The **wavelength**, λ, is the distance between two adjacent peaks or two adjacent troughs.
- The **frequency**, f, of a wave is the number of complete waves produced per second. The unit of frequency is the **hertz (Hz)**. Large frequencies are measured in kilohertz, kHz (10^3 Hz) or megahertz, MHz (10^6 Hz).
- The **period**, T, of a wave is the time taken to produce one wave.

The frequency and time period are linked by the equation:

$$\text{period} = \frac{1}{\text{frequency}}$$

$$T = \frac{1}{f}$$

period, T, in seconds, s
frequency, f, in hertz, Hz

> **Amplitude** is the height of the wave from the middle (undisturbed) position of the string or water.
>
> **Wavelength** is the distance from one point on a wave to the equivalent point on the next wave.
>
> **Frequency** is the number of waves passing a point each second.
>
> **Hertz (Hz)** is the unit of frequency.
>
> **Period** is the time taken to produce one wave – or the time taken for one wave to pass a point.

Example

A wave has a period of 0.14 s. Calculate its frequency.

Answer

$$T = \frac{1}{f}$$

$$0.14 = \frac{1}{f}$$

$$f = 7.1\,\text{Hz}$$

Wave speed and the wave equation

The wave speed is the speed at which energy is transferred (or the wave moves) through a medium.

All waves obey this equation:

wave speed = frequency × wavelength

$$v = f\lambda$$

wave speed, v, in metres per second, m/s

frequency, f, in hertz, Hz

wavelength, λ, in metres, m

Example

Sound travels at 330 m/s in air. Calculate the wavelength of a sound wave having a frequency of 900 Hz.

Answer

$$v = f\lambda$$

$$330 = 900 \times \lambda$$

$$\lambda = \frac{330}{900}$$

$$= 0.37\,\text{m}$$

Now test yourself

TESTED

4 Figure 6.6 shows waves travelling at the same speed on two ropes. Use the correct terms to describe the differences between the waves on rope A and rope B.

rope A   rope B

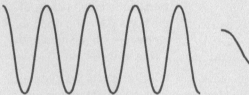

Figure 6.6

5 (a) Two waves have periods of:
    (i) 0.02 s
    (ii) 0.001 s
    Calculate their frequencies.
  (b) Two waves have frequencies of:
    (i) 10^9 Hz (10^9 Hz = 1 gigahertz = 1 GHz)
    (ii) 2 MHz
    Calculate their periods.
6 A stone is thrown into a pond and waves spread out. The speed of the waves is 1.2 m/s and their wavelength 20 cm.
  Calculate the frequency of the waves.

7 Figure 6.7 shows a graph of a wave on a slinky: the *x*-axis shows the distance along the slinky, and the *y*-axis shows the displacement of the slinky from its undisplaced position.

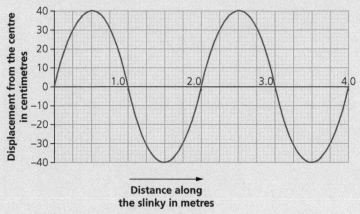

Figure 6.7

(a) What type of wave is moving along the slinky?
(b) (i) State the wave's amplitude.
  (ii) State the wave's wavelength.
(c) The teacher's hand, that produces the waves, moves backwards and forwards from +40 cm to −40 cm and back to +40 cm twice per second.
  (i) State the frequency of the waves.
  (ii) State the time period of the waves.
(d) Calculate the speed of the waves on the slinky.

Answers on p. 149

Measuring wave speeds

REVISED

The speed of sound

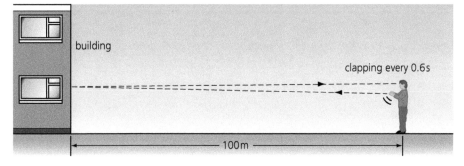

Figure 6.8

- A student claps his hands, and he claps again as he hears the echo from the building.
- His friend finds that he does 10 claps in 6 seconds.
- He calculates the speed of sound.

$$v = \frac{d}{t}$$
$$= \frac{200\,\text{m}}{0.6\,\text{s}}$$
$$= 330\,\text{m/s}$$

Typical mistake

In measuring the speed of sound, remember that the sound has to go to the building **and back**. So the distance travelled here is 200 m not 100 m.

Required practical 8

(a) Investigating waves in a ripple tank

- A ripple tank is set up and the wave generator is attached to a variable frequency supply; the frequency is 12 Hz.
- The pattern seen in Figure 6.9 can be photographed using a mobile phone.
- Now we can work out the wave speed. Since 8 wave crests can be seen in the ripple tank length of 32 cm:

$$\text{wavelength} = \frac{32\,\text{cm}}{8}$$

$$= 4\,\text{cm}$$

$$v = f\lambda$$

$$= 12 \times 4$$

$$= 48\,\text{cm/s}$$

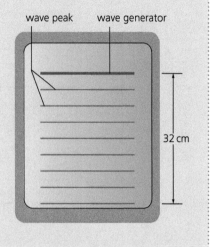

Figure 6.9

(b) Waves in a stretched spring

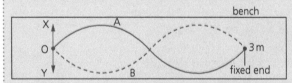

Figure 6.10

- In Figure 6.10 a slinky is stretched along a bench. It is 3.0 m long.
- A student moves one end from X to Y and back to X, so that one wavelength fits into the spring length. Sometimes the shape looks like A, then we see pattern B.
- Another student measures the time for 10 complete oscillations X → Y → X. He finds 10 oscillations take 6.7 seconds.
- Now we can work out the wave speed.

$$\text{wavelength, } \lambda = 3.0\,\text{m}$$

$$\text{period, } T = \frac{6.7}{10}$$

$$= 0.67\,\text{s}$$

$$\text{frequency, } f = \frac{1}{T}$$

$$= \frac{1}{0.67}$$

$$= 1.5\,\text{Hz}$$

$$\text{speed} = \text{frequency} \times \text{wavelength}$$

$$= 1.5 \times 3.0$$

$$= 4.5\,\text{m/s}$$

We can check our result by a second experiment shown in Figure 6.11.

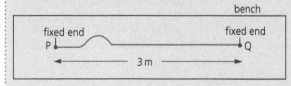

Figure 6.11

- Both ends of the slinky are fixed, and a wave pulse is produced by pulling the slinky to one side and letting it go. The pulse takes 2.7 s to go up and down the slinky twice, a distance of 12 m.
- Now we can calculate the speed as the distance travelled is 12.0 m in 2.7 s:

$$v = \frac{d}{t}$$

$$= \frac{12.0}{2.7}$$

$$= 4.4 \, \text{m/s}$$

Waves travelling from one medium to another

In Figure 6.12 you can see a ripple tank viewed from the top. The black lines represent the peaks of waves.

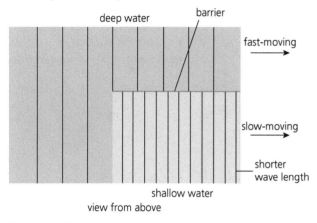

Figure 6.12

Waves travel from left to right, starting in deep water (dark blue), with some waves entering some shallow water (light blue).

The waves travel more slowly in shallow water, but the frequency of the waves stays the same.

Because $v = f\lambda$ the wavelength, λ, reduces in shallow water as the speed, v, reduces.

Now test yourself

8 Describe experiments to measure:
  (a) the speed of sound
  (b) the speed of a wave along a stretched spring.
  In each case, state the measurements you would take and explain how you would use the measurement to calculate the wave speed.
9 Sound waves travel in air at 330 m/s and in carbon dioxide at 270 m/s.
  A sound travels from air into a balloon inflated with carbon dioxide.
  State what happens to:
  (a) the frequency of the sound
  (b) the wavelength of the sound.

Answers on p. 149

Reflection

REVISED

Figure 6.13 shows a light ray being reflected off a mirror. Note the position of the **normal**.

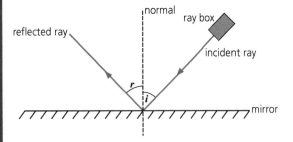

Figure 6.13

The rule of reflection is stated as follows:

the **angle of incidence**, i = the **angle of reflection**, r

The rule of reflection holds for all waves including sound and water waves. Figure 6.14 shows water waves being reflected in a ripple tank.

> The **normal** is a line drawn at 90° to a surface where waves are incident.
>
> The **angle of incidence** is the angle between the incident ray and the normal.
>
> The **angle of reflection** is the angle between the reflected ray and the normal.

> **Exam tip**
>
> Note that water wave fronts are at right angles to the line of travel (as shown by light rays in Figure 6.14).
>
>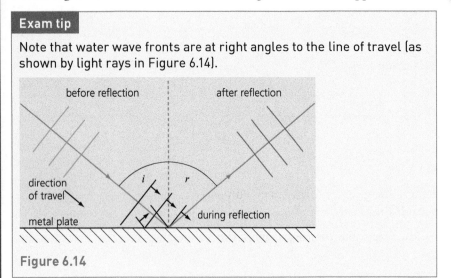
>
> **Figure 6.14**

Transmission and absorption of waves

REVISED

A mirror reflects nearly all the light that is incident on it. Some other surfaces either **transmit** or **absorb** light.

Figure 6.15 shows a light ray incident on a block of glass. Some light is reflected and some is transmitted through the block. You can see that the transmitted ray changes direction.

> A medium that **transmits** light allows the light to pass through it.
>
> A medium that **absorbs** light does not allow the light to pass through it.

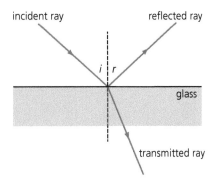

Figure 6.15

Some surfaces absorb light. A room painted with dark paint is darker than a room painted with white paint; the dark paint absorbs light.

The words absorption and transmission refer to all types of waves. For example, some surfaces absorb sound, others allow sound to pass through.

Now test yourself

10 Draw a diagram to show the reflection of water waves off a barrier.
11 Copy the diagrams below and use a protractor to show the paths of the rays after reflection.

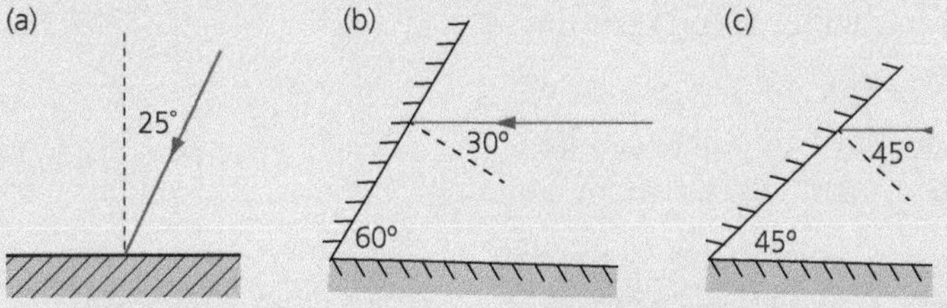

(a) 25°

(b) 30° 60°

(c) 45° 45°

Figure 6.16

12 (a) Explain the meanings of the words: *absorption* and *transmission*.
   (b) Give an example of a material that:
      (i) reflects sound
      (ii) transmits sound
      (iii) absorbs sound.

Answers on p. 149

ℍ Sound waves

Sound waves can travel through gases, liquids and solids. Figure 6.17 shows a tuning fork vibrating; as it vibrates it sets up compressions and rarefactions in air – these are sound waves.

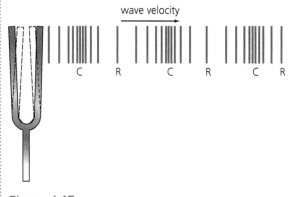

wave velocity

C   R   C   R   C   R

Figure 6.17

The compressions and rarefactions also set our eardrums in motion. The vibrations of the ear drum, in turn, set up vibrations in bones, which is how we detect sound. The conversion of sound waves to vibrations of solids works over a limited frequency range. This restricts the limits of human hearing.

We can hear sounds in the frequency range 20 Hz–20 000 Hz.

Ultrasound

Ultrasound waves are sound waves with a frequency above 20 kHz; these frequencies are outside our hearing range. Very high frequency ultrasound waves are used to scan our bodies.

- At each surface, some waves are reflected and some transmitted. A computer can build up a picture, using the intensity and timing of the reflected waves (echoes).
- Ultrasound scanning is safe for patients and is used for foetal scanning.

Ships use narrow beams of ultrasound to look for fish, for navigation and naval ships can look for submarines. The depth of an object can be determined from the time it takes for the echo of sound to return.

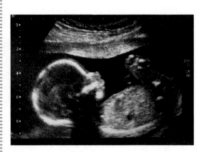

Figure 6.18 **Ultrasound scans are used for checking on babies in the womb. This is called foetal scanning.**

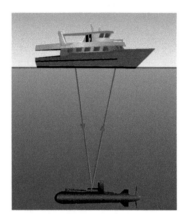

Figure 6.19

Seismic waves

Seismic waves are produced by earthquakes. Such waves transmit energy and also give us information about the structure of the Earth.

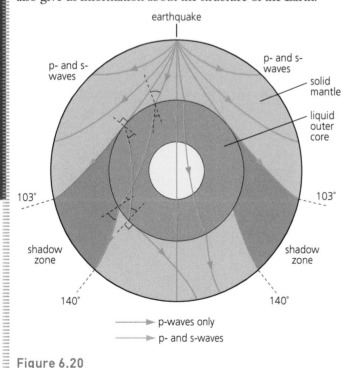

Figure 6.20

Seismic waves consist of **p-waves** and **s-waves**.

- p-waves are longitudinal seismic waves. These waves travel at different speeds through solid rock and liquid rock.
- s-waves are transverse seismic waves. These waves can only travel through solid rock, they cannot travel through liquid rock.

The fact that s-waves cannot travel through liquid rock allowed scientists to determine:

- the Earth has a solid outer layer – the mantle
- the Earth has a liquid outer core
- further measurements show that the Earth has a solid inner core.

> **p-waves** are longitudinal seismic waves.
>
> **s-waves** are transverse seismic waves.

Now test yourself

TESTED ☐

13 (a) What is ultrasound?
   (b) Explain the uses of ultrasound.
14 Ultrasound is used to investigate a patient's kidney. Ultrasound has a speed of 1600 m/s in the body, and a frequency of 4 MHz is used.
   Calculate the wavelength of the ultrasound.
15 (a) Explain the difference between seismic p-waves and seismic s-waves.
   (b) Explain how s-waves and p-waves allowed scientists to discover that the Earth has a liquid core.
16 The ship in Figure 6.19 is using ultrasound with a frequency of 25 kHz. The transmitted ultrasound travels with a speed of 1500 m/s in water.
   (a) Calculate the wavelength of the ultrasound.
   (b) (i) A pulse of ultrasound is emitted and an echo from the submarine is detected 0.18 s later. Calculate the depth of the submarine.
       (ii) Half a minute later, the pulses are detected 0.24 s after being emitted. Calculate the rate at which the submarine is diving.
   (c) Modern submarines are coated with rubber 4 mm thick. The rubber is full of small holes of varying size. Explain the purpose of the rubber.

Answers on pp. 149–150

Electromagnetic waves

REVISED ☐

Electromagnetic waves are transverse waves that transfer energy from the source to an absorber.

Electromagnetic waves form a continuous spectrum, with wavelength varying from 10^{-12} m to over 1 km. Electromagnetic waves are grouped by wavelength. These groups are: radio, microwave, infrared, visible light, ultraviolet, x-rays and gamma rays.

Electromagnetic waves all have these properties.

- They are transverse waves.
- They transfer energy from one place to another.
- They obey the equation $v = f\lambda$.
- They travel through a vacuum.
- They all travel at the same speed in a vacuum – 300 000 000 m/s (3×10^8 m/s).

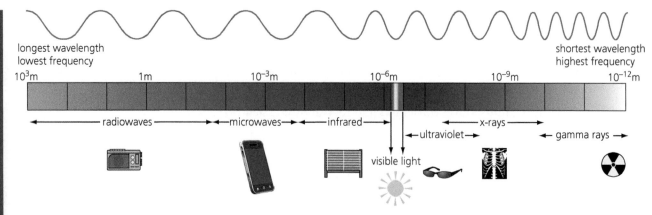

Figure 6.21

Properties of electromagnetic waves 1

Different substances absorb, transmit, refract or reflect electromagnetic waves in ways that vary with wavelength.

- We see different colours – a green shirt reflects green light, but absorbs all other colours.
- A polished metal surface reflects wavelengths of electromagnetic waves from radio waves to ultraviolet, but x-rays and gamma rays are transmitted by thin metal plates.
- Food is cooked in a microwave oven. The wavelength of microwaves is carefully chosen so that they are absorbed by water.

Refraction

Figure 6.22 shows the **refraction** of a light ray as it passes through a glass block.

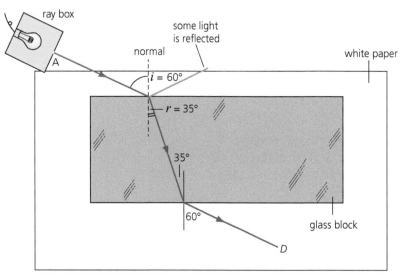

> When a wave is transmitted from one medium to another, the transmitted wave changes direction. This is called **refraction**. All types of waves, including light and sound, refract when they travel from one medium to another.

Figure 6.22

- Light is refracted when it enters and leaves glass (or water) because the speed of light is greater in air than it is in glass (or water).
- Light bends towards the normal when it enters glass, and away from the normal when it travels from glass into air.

All waves show refraction.

- In Figure 6.20 you can see that seismic waves refract as they travel from the mantle to the outer core; they bend towards the normal because the waves travel faster in the mantle than the core.
- Water waves are refracted when they travel from deep water to shallow water, as shown in Figure 6.23.

Here the waves:
- slow down
- become shorter in wavelength
- change direction.

> **Exam tip**
>
> Note that the waves do not refract (change direction) when they travel parallel to the normal.

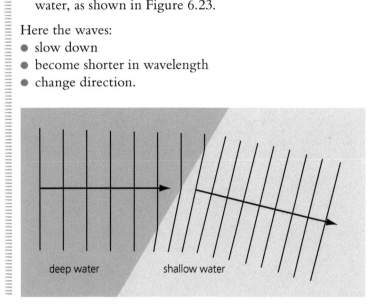

Figure 6.23

Now test yourself

17 The speed of electromagnetic waves is 3×10^8 m/s.

(a) Red light has a wavelength of 6.5×10^{-7} m. Calculate the frequency of red light.

(b) A radio wave has a frequency of 2 MHz. Calculate its wavelength.

18 Draw diagrams to show how the water waves are refracted in each of the following diagrams.

> **Exam tip**
>
> It is easier to use standard form to solve problems with very large or very small numbers.

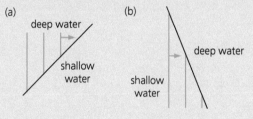

Figure 6.24

19 Draw diagrams to show how light is refracted in each of the following cases.

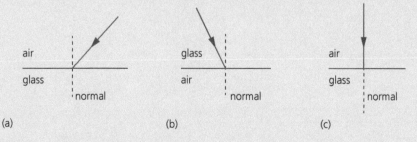

Figure 6.25

20 Refer to Figure 6.20. Explain how the green rays show that p-waves travel more slowly in the outer core than they do in the mantle.

Answers on p. 150

Required practical 9

Investigating the reflection of light 9a

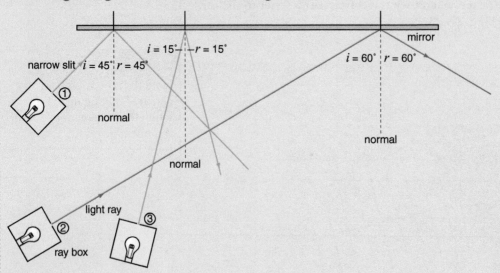

Figure 6.26

Figure 6.26 shows how you can investigate the reflection of light.
1 Using a support, set a plane mirror upright on a large piece of paper.
2 Draw a line along the front of the mirror.
3 Use a ray box to shine an incident ray on the mirror.
4 Mark the positions of the incident and reflected rays.
5 Remove the mirror and carefully draw in the incident and reflected rays, and the normal.
6 Use a protractor to measure the angles i and r.
7 Analyse the results to find the relationship between i and r.

Investigating the refraction of light 9b

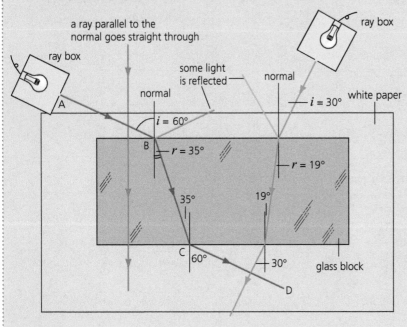

Figure 6.27

Figure 6.27 shows how you can investigate the refraction of light.
1 Place a block of glass onto a large piece of paper and draw round the block.
2 Shine a ray at the glass, e.g. ray AB in Figure 6.27. This is the incident ray.
3 Carefully mark on the paper the rays AB and CD.
4 Then lift up the block and draw in the two normals and the ray BC.
5 Now measure the angle of incidence *i* and the angle of refraction *r*.
6 Repeat for different angles.
7 The table shows some typical results.
   Draw a graph of the angle of incidence against the angle of refraction.

Angle of incidence, *i*	Angle of refraction, *r*
0	0
23°	15°
34°	22°
48°	30°
59°	35°
80°	41°

⊕ Properties of electromagnetic waves 2

REVISED

Figure 6.28 shows how radio waves and microwaves are transmitted and received.

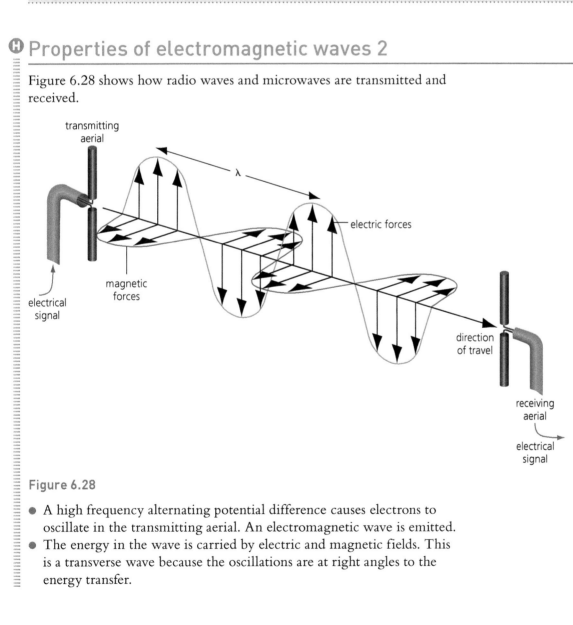

Figure 6.28

- A high frequency alternating potential difference causes electrons to oscillate in the transmitting aerial. An electromagnetic wave is emitted.
- The energy in the wave is carried by electric and magnetic fields. This is a transverse wave because the oscillations are at right angles to the energy transfer.

- The radio waves are absorbed by the receiving aerial. They create an alternating current with the same frequency as the radio wave itself. So radio waves induce oscillations in an electrical circuit.
- Radio waves also carry information that we hear and see on televisions.
- Electromagnetic waves of wavelengths in the infrared to x-ray range are produced by changes in atoms. Gamma rays originate from changes in the nucleus of an atom.

Hazards of radiation

Ultraviolet waves, x-rays and gamma rays can be hazardous to humans.
- Ultraviolet waves can cause skin to age prematurely and increase the risk of skin cancer. Ultraviolet waves from the Sun cause sun tanning.
- X-rays and gamma rays are ionising radiations that can cause the mutation of genes and cancer. Radiation doses are measured in sieverts; the dose is a measure of the risk to us (see Chapter 4 page 54).
- Sunburn is caused by infrared radiation.

Uses and applications of electromagnetic waves

REVISED

Electromagnetic waves have many practical applications. For example:
- radiowaves – television and radio
- microwaves – satellite communications, cooking food
- infrared – electrical heaters, cooking food, infrared cameras
- visible light – sight, fibre optic communications
- ultraviolet – energy efficient lamps
- X-rays and gamma rays – medical imaging and treatments.
- Radio waves are used to transmit radio and television signals. Long wavelengths are suitable for transmissions around the Earth's surface.
- Microwaves are used for satellite communications and cooking. A narrow beam of microwaves is suitable for directing towards satellites. Some microwaves have the correct wavelength to be absorbed by water, thus allowing food to cook.
- Infrared radiation produces a heating effect, so we use infrared waves in electric heaters and in ovens for cooking. Warm objects also emit infrared waves; so an infrared camera allows us to see things at night. Infrared waves are also used in remote controls.
- We use visible light all the time to see. Light (or infrared) is also used in fibre optic communications. Fibre optic links are widely used in telecommunications.
- Some substances can absorb ultraviolet radiation and then emit the energy as visible light. This is called fluorescence. This principle is used in some energy efficient lamps. Fluorescence also has applications in crime solving. Possessions can be marked with an invisible fluorescent spray, which can be seen in ultraviolet light.
- X-rays (and gamma rays) can penetrate our bodies and can be used in medical imaging.
- Gamma rays (and x-rays) can be used in medical treatment. These radiations can be hazardous to our bodies, but they can also be used effectively to destroy cancerous tissue.

Now test yourself

21 (a) Name two parts of the electromagnetic spectrum that are hazardous to humans.
(b) For each of your choices explain what the hazards are.
22 Give and explain a use for:
(a) microwaves
(b) infrared waves.
23 List the seven types of wave in the electromagnetic spectrum starting with the waves with the lowest frequency.

Answers on p. 150

Required practical 10

Investigating the emission of infrared radiation from different surfaces

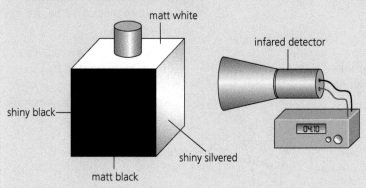

Figure 6.29

The apparatus shows a 'Leslie Cube'. It is filled with hot water. The infrared detector is placed close to each face to detect how much radiation is emitted from each surface. We make it a fair test by keeping the detector the same distance from each face.

This investigation shows us that:
● dull black surfaces are good emitters of infrared radiation
● shiny or white surfaces are poor emitters of infrared radiation.

We can also investigate which type of surfaces are good absorbers of radiation.

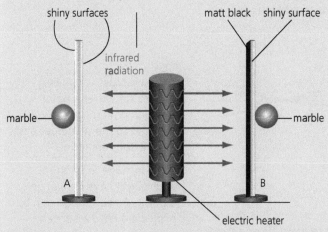

Figure 6.30

In Figure 6.30 two marbles are stuck with wax onto two metal sheets. One sheet (B) has a dull black surface facing the electric heater; the other sheet (A) has a shiny metallic surface facing the heater.

The marble falls quickly from side B, and remains on side A.
● Dull black surfaces are good absorbers of infrared radiation.
● Shiny or white surfaces are poor absorbers of infrared radiation.

Now test yourself

24 In the experiment with the marbles (Figure 6.30), explain what measures must be taken to make it a fair test.
25 Below are four types of surface.

dull black   shiny black   dull white   shiny metallic

Organise the list in the order of:
(a) the best emitters of infrared radiation
(b) the best absorbers of infrared radiation.

Answers on p. 150

Refraction in lenses

A lens forms an image by refracting light.

There are two types of lens:
● a **convex** (or converging) **lens**
● a **concave** (or diverging) **lens**.

Figure 6.31 shows the refraction in each lens. Light bends towards the normal when entering the lens, and away from the normal on leaving the lens.

> In ray diagrams a **convex lens** is represented by the symbol ⤷
>
> In ray diagrams a **concave lens** is represented by the symbol ⟩

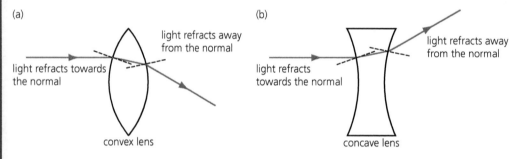

(a) **light refracts away from the normal**
light refracts towards the normal
convex lens

(b) **light refracts away from the normal**
light refracts towards the normal
concave lens

Figure 6.31

Convex lens

Figure 6.32 shows parallel rays of light being focused by a convex lens. The rays converge on a point called the **principal focus**. There is a principal focus on both sides of the lens.

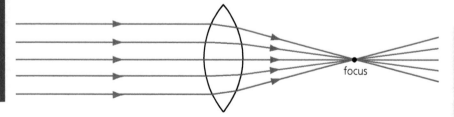

focus

Figure 6.32

> The **principal focus** of a convex lens is the point through which light rays parallel to the principal axis pass after refraction.
>
> A **real image** is formed when light rays converge to a point. A real image can be projected onto a screen.

Ray diagrams

In Figure 6.33 a convex lens focuses rays from an object to form a **real image**.

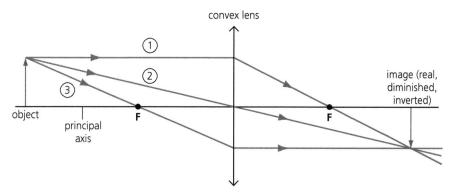

Figure 6.33

Rays come to the lens from all angles, but there are three rays which can be used to locate the image. We can predict the paths of these rays through the lens.

- Ray 1: a ray parallel to the principal axis (on the left) passes through the principal focus on the right.
- Ray 2: a ray passing through the centre of the lens does not change direction.
- Ray 3: a ray passing through the principal focus on the left, emerges from the lens parallel to the principal axis on the right.

We describe this image as:
- real (because the rays meet)
- inverted (upside down)
- diminished (smaller than the object).

We can find the image by using only two rays as shown in Figure 6.34.

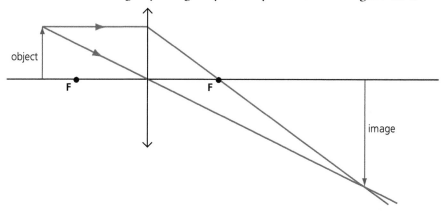

> **Exam tip**
>
> Refraction of light takes place at both surfaces of a lens. But in drawing ray diagrams we make it easier by showing the refraction occurring at the thin line shown by the symbol ↕

Figure 6.34

In this case the image is: real, inverted and magnified – because it is larger than the object.

The magnification is given by the equation:

$$\text{magnification} = \frac{\text{image height}}{\text{object height}}$$

$$= \frac{28\,\text{mm}}{14\,\text{mm}}$$

$$= 2$$

A magnifying glass

Figure 6.35 shows how we can use a convex lens as a magnifying glass.

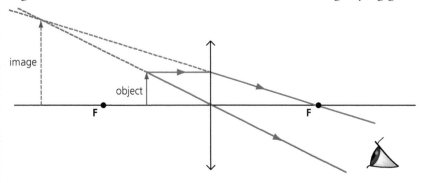

Figure 6.35

An object is placed inside the **focal length** of the lens. Now when the rays are refracted, they do not meet. When an eye looks at these rays, they appear to come from a point further behind the lens and the object appears magnified.

The image is:
- virtual
- upright (the right way up)
- magnified.

Concave lenses

Figure 6.36 shows the refraction of light by a concave (diverging) lens. Rays parallel to the principal axis diverge as if they have come from a point behind the lens. This point is the principal focus of the lens, but it is a **virtual focus**, because the rays appear to come from it.

> The **focal length** of a lens is the distance between the lens and the principal focus.
>
> A **virtual image** can only be seen by one person. The rays appear to diverge from the image. The image cannot be seen on a screen.

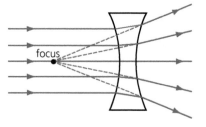

Figure 6.36

Images and ray diagrams

Figure 6.37 shows how we draw a ray diagram for a concave lens.

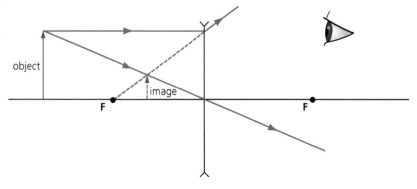

Figure 6.37

One ray passes through the centre of the lens without changing direction.

A second ray parallel to the principal axis diverges as if it has come from the principal focus. An eye looking at these rays sees a virtual image.

Wherever the object is the image seen through a concave lens is always:
- virtual
- upright
- diminished.

Now test yourself

26 (a) Explain what is meant by:
   (i) a real image
   (ii) a virtual image.
   (b) Which type of lens only produces a virtual image?
27 Figure 6.38 shows four different examples of ray diagrams. Copy the diagrams onto graph paper.
   (a) In each case, complete the diagrams to show how the rays can be extended to find the image position.
   (b) In each case, state whether the image is real or virtual.
   (c) For each case, calculate the magnification of the image.

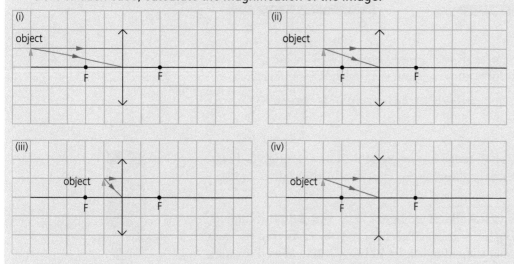

Figure 6.38

28 A convex lens produces a magnification of ×2. The object is placed 6 cm from the lens.
   (a) Draw a scale diagram to show the position of the object and lens.
   (b) Draw rays through the lens to locate the position of the image, which has a magnification of ×2.
   (c) Use the diagram to determine the focal length of the lens.

Answers on pp. 150–151

Visible light

Each colour within the visible spectrum has its own narrow band of wavelength and frequency.

Figure 6.39 shows how the colours of the rainbow are related to their wavelengths.

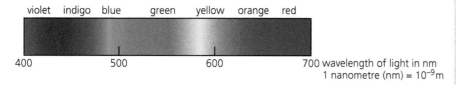

violet   indigo   blue   green   yellow   orange   red

400   500   600   700 wavelength of light in nm
                                           1 nanometre (nm) = 10^{-9} m

Figure 6.39

Reflection

Reflection from a smooth surface (such as a mirror) in a single direction is called **specular reflection**. Reflection from a rough surface causes scattering: this is called **diffuse reflection**.

The colour of an **opaque** object is determined by which wavelengths of light are more strongly reflected.

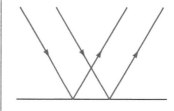

Figure 6.40 This is **specular reflection**.

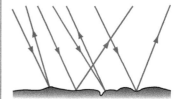

Figure 6.41 This is **diffuse reflection**.

An **opaque** object does not allow light to pass through it.

For example, in Figure 6.42 the object looks blue because only blue light is reflected. All other colours are absorbed. Similarly, a red object looks red because it reflects red light and absorbs all other colours.

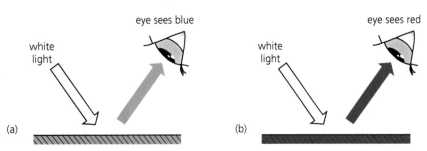

Figure 6.42

- An object which reflects all wavelengths equally looks white.
- An object which absorbs all wavelengths completely looks black.

Objects that transmit light are either **transparent** or **translucent**.

Coloured filters

A filter only allows a small range of wavelengths to pass through it. For example, a green filter only allows green light to pass through it, and it absorbs all other colours.

In Figure 6.43, the red filter allows red light to pass through. But the red light cannot pass through the green filter. So a red object viewed through a red filter looks red, but the red object looks black when viewed through a blue or green filter.

> A **transparent** object allows us to see clearly through it. Glass is transparent.
>
> A **translucent** object allows light to pass through, but we cannot see objects clearly through it. Some plastics and frosted glass are translucent.

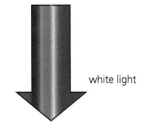
white light

red filter

red light

green filter
no light

Figure 6.43

Now test yourself

TESTED ☐

29 What colour does a green book look when viewed through:
  (a) a blue filter
  (b) a green filter
  (c) a red filter?
30 A boy goes to a disco wearing green trousers, a red shirt and a blue cap.
  What colour do his clothes look in red light?
31 Explain how the reflection and absorption of different wavelengths of light cause the colours we see.
32 Light travels at 3×10^8 m/s.
  (a) Calculate the frequency of blue light with a wavelength of 4.8×10^{-7} m.
  (b) Calculate the wavelength of orange light with a frequency of 5.0×10^{14} Hz.

Answers on p. 151

Black body radiation

REVISED ☐

Emission and absorption of infrared radiation

You can feel infrared radiation when the Sun shines on a warm day, or you feel the radiation when you put your hand in a warm oven.
- All objects emit and absorb infrared radiation, no matter what their temperature.
- The hotter the object, the more radiation is emitted per second.
- At higher temperatures some energy can also be emitted as visible light.
- The shorter the wavelength of infrared radiation (or visible light) emitted from an object, the hotter its temperature.

Exam practice answers and quick quizzes at **www.hoddereducation.co.uk/myrevisionnotes**

A **perfect black body** is a perfect emitter and absorber of radiation.

A **perfect black body** absorbs all the radiation incident on it. A perfect black body is also the best possible emitter.

ⓗ Radiation and temperature change

- An object at a constant temperature is absorbing radiation at the same rate as it is emitting it (assuming there are no other energy transfer processes).
- The temperature of an object increases when it absorbs radiation faster than it emits radiation.
- The temperature of an object decreases when it emits radiation faster than it absorbs radiation.

The temperature of the Earth depends on the rates of emission and absorption of radiation, and the reflection of radiation into space.

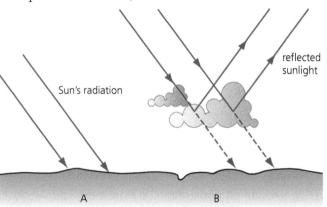

Figure 6.44

In Figure 6.44 position A is hotter than position B, because above position B, the clouds reflect some of the radiation before it reaches the Earth's surface. This means position A warms more quickly when the Sun rises.

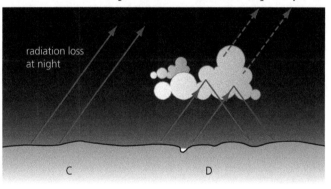

Figure 6.45

In Figure 6.45 it is night. At night the Earth cools because radiation is emitted and not absorbed. Now the clouds keep us warm. At position C the Earth cools quickly as radiation is emitted. Position D stays warm for longer because the clouds absorb some of the emitted radiation.

Now test yourself

TESTED ☐

33 A joint of meat is put into the oven to cook.
  (a) For the first half hour the temperature of the meat rises. Explain why.
  (b) For the next hour the meat reaches a constant temperature. Explain why.
  (c) When the meat is removed from the oven, its temperature falls. Explain why.
34 Explain what is meant by a perfect black body.

Answers on p. 151

Summary

- Waves transfer energy and information.
- In a transverse wave the vibrations of the wave are at right angles to the direction of energy transfer.
- In a longitudinal wave the vibrations are parallel to the direction of energy transfer.
- Amplitude, A, is the height of a wave measured from its undisturbed position.
- Wavelength, λ, is the distance between two adjacent peaks (or troughs).
- Frequency, f, is the number of waves produced per second.
- Period, T, is the time taken to produce one wave.
- $f = \dfrac{1}{T}$
- $v = f\lambda$
- Reflection: the angle of incidence, i, = the angle of reflection, r.
- The angles i and r are measured between the rays and the normal.
- The normal is a line perpendicular to the surface of the mirror.
- Sound is a longitudinal wave.
- Ultrasound is a sound wave with a frequency above the human range of hearing (20 Hz to 20 000 Hz).
- A p-wave is a longitudinal seismic wave.
- An s-wave is a transverse seismic wave (which can only travel through solid rock).
- There are seven types of electromagnetic wave: radio, microwaves, infrared, visible light, ultraviolet, x-rays and gamma rays.

- Electromagnetic waves travel at the speed of light: 3×10^8 m/s (in a vacuum).
- Electromagnetic waves are transverse waves.
- Refraction is the name given to the change of direction when a wave travels from one material to another.
- Light travels faster in air than in glass.
- When a light ray enters glass from air, it bends towards the normal. The ray bends away from the normal when it travels from glass to air.
- A converging lens brings light to a focus.
- A converging lens can produce real or virtual images.
- A diverging lens can only produce a virtual image.
- $\text{magnification} = \dfrac{\text{image height}}{\text{object height}}$
- Specular reflection occurs off smooth surfaces.
- Diffuse reflection occurs when light is scattered off rough surfaces.
- The colour of an object is determined by the wavelength of light the object reflects.
- A filter allows a particular colour of light to pass through it.
- All objects emit and absorb infrared radiation.
- A black body is a perfect absorber and emitter of radiation.

Exam practice

1 (a) Light from an object forms an image in a plane mirror.
   State which two statements below are correct. [2]
   1 The image in the plane mirror is virtual.
   2 Light from the object passes through the image.
   3 Light waves are longitudinal.
   4 The angle of incidence is always equal to the angle of reflection.
   5 The incident and reflected rays are always at right angles to each other.

(b) (i) Copy Figure 6.46 and add the correct labels to the numbers in the diagram. [2]
   (ii) Write *r* on your diagram to show the angle of reflection. [1]

(c) Figure 6.47 shows a ray of light travelling from air to glass.
   (i) Copy the diagram and add labels to show the angle of incidence, *i*, and the angle of refraction, *r*. [2]
   (ii) Describe how you would determine the path of another ray passing through the block, which has a different angle of incidence. You should include a description of the apparatus you would use. [4]

2 Figure 6.48 shows transverse waves on a piece of thick string. A person produces the waves by holding the string and moving it from side to side. The end A travels through 5 complete movements (from side to side and back) in 2 seconds.

(a) Explain what is meant by a transverse wave. [2]

(b) Use the diagram to calculate:
   (i) the amplitude of the waves [1]
   (ii) the wavelength of the waves. [1]

(c) Calculate:
   (i) the frequency of the waves [1]
   (ii) the period of the waves. [1]

(d) Calculate the speed of the waves. [3]

3 Figure 6.49 shows water waves in a ripple tank. The waves are travelling in deep water.
   The waves cross a boundary into shallow water, where they travel more slowly.
   Copy the diagram and show what happens to the waves when they travel in the shallow water. [3]

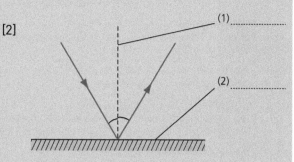

(1) ....
(2) ....

Figure 6.46

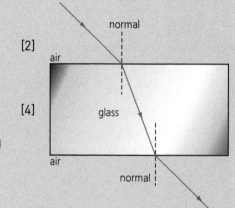

normal

air

glass

air

normal

Figure 6.47

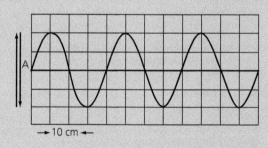

A

← 10 cm →

Figure 6.48

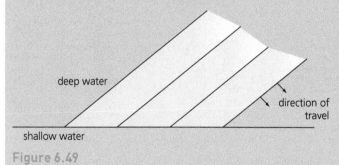

deep water

direction of travel

shallow water

Figure 6.49

4 Figure 6.50 shows how the depth of the oil in a storage tank is measured using sound of frequency 15 kHz. The sound is emitted and received by a transducer at the bottom of the tank. Sound is reflected off the surface of the oil. The time interval between the transmitted and reflected waves is displayed on the oscilloscope.

(a) State the range of frequencies that a human can hear. [1]

(b) The speed of sound in the oil is 1200 m/s. Calculate the wavelength of the waves, using the information above. [3]

(c) Use the information on the oscilloscope trace to show that the time taken to travel from the transducer to the surface and back is 8 ms. [1]

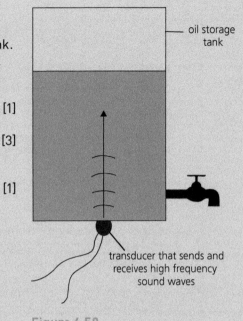

Figure 6.50

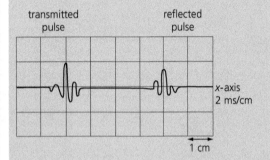

Figure 6.51

(d) Calculate the depth of the liquid. [3]

5 The diagram below represents the electromagnetic spectrum

A	microwaves	infrared	light	B	X-rays	gamma rays

(a) Name the parts of the spectrum labelled A and B. [2]
(b) Which electromagnetic radiation is emitted by a hot oven? [1]
(c) Which electromagnetic radiation is used for communication with satellites? [1]
(d) An X-ray has a wavelength of 1.2×10^{-10} m.
Calculate the frequency of the wave. The speed of light is 3.0×10^{8} m/s. [3]

6 Figure 6.52 shows rays passing through a lens and an image being formed.

(a) State whether the lens is convex or concave. [1]
(b) Use three words to describe the image. [3]
(c) Calculate the magnification of the image. [2]

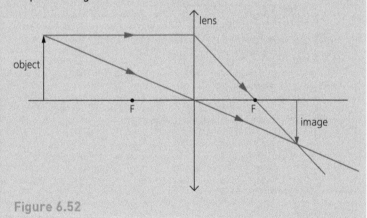

Figure 6.52

7 Figure 6.53 shows two rays from the top of an object. Copy the diagram.
  (a) Show how the two rays pass through the lens. [2]
  (b) Now determine the position of the image. Show this on your diagram. [2]
  (c) Is the image real or virtual? Explain your
      answer. [2]
8 In Figure 6.54 seismic p-waves are travelling
  through the Earth's solid mantle. They refract
  when they meet the Earth's liquid core.
  (a) The p-wave is a longitudinal wave.
      Describe how energy is transferred by
      a longitudinal wave. [2]
  (b) In the mantle the time period of the wave
      is 7.7 s.
      (i) Calculate the frequency of the wave. [2]
      (ii) State what the time period of the wave is
           once it is in the core. [1]
  (c) Calculate the wave speed of the p-waves in the
      mantle, using the information in the diagram. [2]
  (d) (i) Explain the meaning of the word *refraction*. [1]
      (ii) Explain how you can tell from the diagram
           that the p-waves travel more slowly in the core. [2]
      (iii) Use the diagram to estimate the wavelength
            of the p-waves in the core. [1]
      (iv) Use your answer to part (iii) to calculate the
           speed of the p-waves in the core. [2]
9 Design an experiment to measure the speed of sound
  through air. Explain the measurements you would
  take, and how you would ensure that your conclusion
  is accurate.

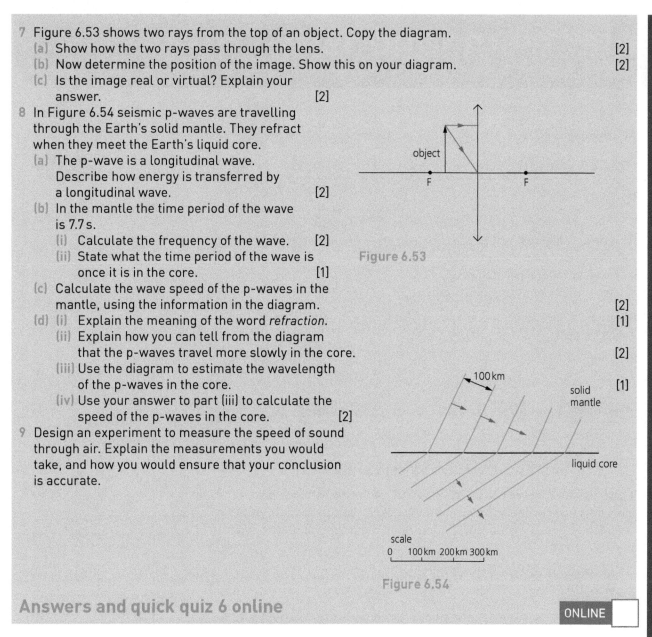

Figure 6.53

Figure 6.54

Answers and quick quiz 6 online

ONLINE

7 Magnetism and electromagnetism

Permanent and induced magnetism, magnetic forces and fields

Poles

The poles of a magnet are the places where the **magnetic** forces are strongest. When two magnets are brought close to each other they exert a force on each other.

- Two like poles repel each other.
- Two unlike poles attract each other.

Magnetic forces of attraction and repulsion are examples of non-contact forces.

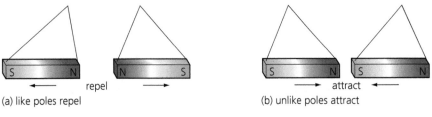

(a) like poles repel   (b) unlike poles attract

Figure 7.1 Two like poles repel, two unlike poles attract.

There are two types of pole, north pole and south pole. A north pole is short for a **north-seeking pole**, and south pole is short for a **south-seeking pole**. If a magnet is suspended, it points along the north–south direction.

> **Magnetic** materials are attracted by a magnet.
>
> A **north-seeking pole** of a magnet points north.
>
> A **south-seeking pole** of a magnet points south.

Permanent and induced magnets

A **permanent magnet** produces its own magnetic field.

A permanent magnet always has a north and a south pole. If you have two permanent magnets, you will be able to show that they can repel each other, as well as attract.

An **induced magnet** is a material that becomes magnetic when it is placed in a magnetic field. It loses its magnetism when it is removed from a magnetic field. An induced magnet is always attracted to a permanent magnet, because the induced magnet is magnetised in the direction of the magnetic field.

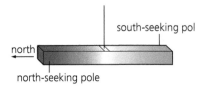

Figure 7.2 The north-seeking pole points towards north, and the south-seeking pole points towards south.

> A **permanent magnet** produces its own magnetic field.
>
> An **induced magnet** becomes a magnet when it is placed in a magnetic field.

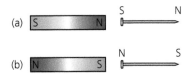

Figure 7.3 The nail becomes magnetised in the direction of the permanent magnet's field. The nail is always attracted to the permanent magnet.

Magnetic fields

The region around a magnet, where a force acts on another magnet or a magnetic material, is called a magnetic field.

The strength of the magnetic field depends on the distance from the magnet. The field is strongest near the poles of the magnet. We use magnetic field lines to represent a magnetic field. Magnetic field lines always start at a north pole and finish at a south pole. When the lines are close together, the field is strong. The further apart the lines are, the weaker the field is.

The direction of a magnetic field can be found using a small plotting compass. The compass needle always points along the direction of the field, as shown in Figure 7.4

> **Exam tip**
>
> The most common magnetic materials are iron, steel, cobalt and nickel.

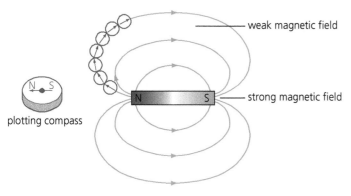

plotting compass

weak magnetic field

strong magnetic field

Figure 7.4

The Earth's magnetic field

We use a compass to help us navigate. The Earth has a magnetic field. The north (seeking) pole of the magnet points towards magnetic north. Figure 7.5 shows the shape of the Earth's field. The north pole of the compass is attracted towards a south (seeking) pole at magnetic north.

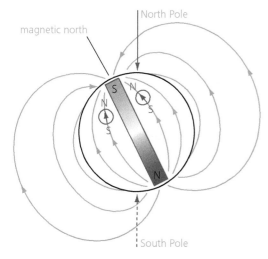

Figure 7.5 The shape of the Earth's magnetic field.

> **Typical mistake**
>
> A careless mistake is to draw two magnetic field lines crossing each other. This cannot be right as it would mean a compass has to point two ways at once. Field lines **never** cross.

Now test yourself

TESTED

1 Name four magnetic materials.
2 How many types of magnetic pole are there? State the rule about the attraction and repulsion of magnetic poles.
3 (a) Explain what is meant by a *permanent magnet*.
   (b) Explain how you can show a magnet is a permanent magnet.
4 Draw the shape and direction of a magnetic field around the bar magnet.
5 What type of pole is there at the magnetic north?
6 In Figure 7.6, three steel paper clips are attached to a magnet.
   (a) The size of the magnetic force is greater than which other force acting on each paper clip?
   (b) (i) What type of magnets are the paper clips?
       (ii) Draw a diagram to show the magnetic poles on each paper clip.
   (c) Explain why the paper clips could be suspended from the south pole of the magnet.
7 Explain why the steel pins repel each other in Figure 7.7.

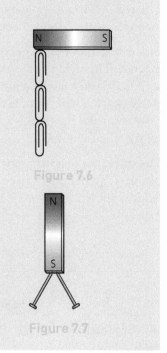

Figure 7.6

Figure 7.7

Answers on p. 151

Electromagnetism

REVISED

When a current flows through a conducting wire, a magnetic field is produced around the wire. The strength of the field:
● depends on the size of the current through the wire
● is weaker further away from the wire.

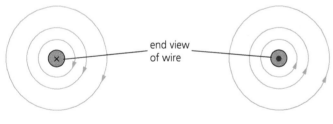

Figure 7.8

Figure 7.8 shows the pattern of magnetic field lines surrounding a wire. When the current flows into the paper (shown as ⊗) the field lines are clockwise. When the current flows out of the paper (shown as ⊙) the field lines are anticlockwise.

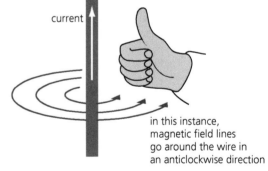

Figure 7.9 **The right-hand grip rule. When you put your thumb along the direction of the current in a wire, your fingers point in the direction of the magnetic field around the wire.**

The field of a solenoid

Figure 7.10 shows the magnetic field that is produced by a current flowing through a long coil of wire (a **solenoid**). The magnetic field has a similar shape to that of a bar magnet.

A solenoid's magnetic field can be increased by:
● increasing the current
● using more turns of wire
● putting the turns closer together
● putting an iron core in the middle of the solenoid.

A **solenoid** is a long coil of wire.

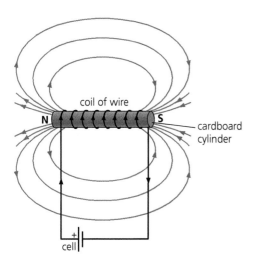

Figure 7.10 The magnetic field around a solenoid has a similar shape to that of a bar magnet.

Electromagnets

Figure 7.11 shows an electromagnet in action. The iron core is an induced magnet, which is magnetised when a current flows through the coils. When the current is switched off, the magnet loses its magnetism and the iron filings fall off.

Figure 7.11

Now test yourself

TESTED

8 List four ways of increasing the strength of the magnetic field produced by a solenoid.

9 Figure 7.12 shows a wire placed vertically with a current flowing into the paper. Copy the diagram, and add lines to show the direction and strength of the magnetic field around the wire.

Figure 7.12

10 Explain how to use the right-hand grip rule, to find the direction of the magnetic field around a wire that carries a current.

11 Sketch the shape of the magnetic field around a solenoid with a current flowing through it.

12 Figure 7.13 shows a solenoid wrapped on a hollow tube.

   (a) State the directions of each of the compass needles 1–6, e.g. pointing to the right, left, up or down.

   (b) Which end of the solenoid acts as a north pole?

   (c) State what happens to the compass needles when the current is reversed.

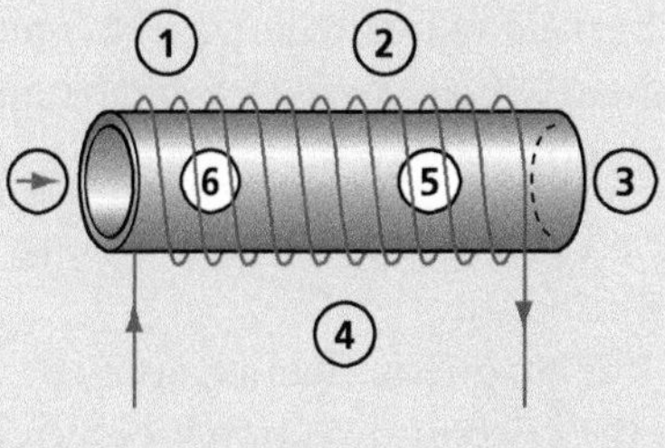

Answers on p. 151

Figure 7.13

ⒽFleming's left-hand rule

When a conductor carrying a current is placed in a magnetic field, the magnet producing the field and the conductor exert a force on each other. This is called the motor effect.

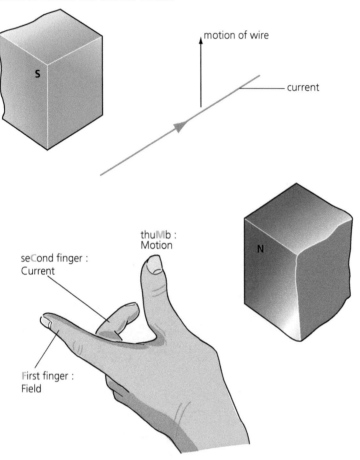

Figure 7.14

The size of the force on the wire depends on:
- the strength of the magnetic field from the magnet
- the size of the current
- the length of the wire between the poles of the magnet.

The left-hand rule allows us to predict the direction of the force on the wire. You arrange your thumb, first finger and second finger so that they are at right angles to each other, as shown in Figure 7.14.
- The first finger points in the direction of the magnetic field – north to south.
- The second finger points in the direction of the current.
- Then the thumb points along the direction of the force that causes the wire to move.

This rule works when the current and field are at right angles to each other. When the field and current are parallel to each other, the force on the wire is zero.

You can use the left-hand rule to predict that the direction of the force is reversed if:
- the magnetic field is reversed
- the current direction is reversed.

⊕ Magnetic flux density

We represent magnetic fields by drawing lines that show the direction of a force on a north pole. These field lines are formally known as **lines of magnetic flux**.

Figure 7.15 shows the lines of flux between two pairs of magnets. The magnets in Figure 7.15 (b) are stronger than the magnets in Figure 7.15 (a), so they exert a stronger force on a magnetic material.

We show stronger magnets by drawing more lines of flux in a given area.

The strength of the magnetic force is determined by the flux density, B.

> Flux density, B, is the number of lines of flux in a given area. (Flux density is sometimes called the B-field.)

Calculating the force

The force on a wire of length, L, carrying a current, I, at right angles to a magnetic field is given by the equation:

force = magnetic flux density × current × length

$$F = BIL$$

> force, F, in newtons, N
>
> magnetic flux density, B, in tesla, T
>
> current, I, in amperes (amps), A
>
> length, L, in metres, m

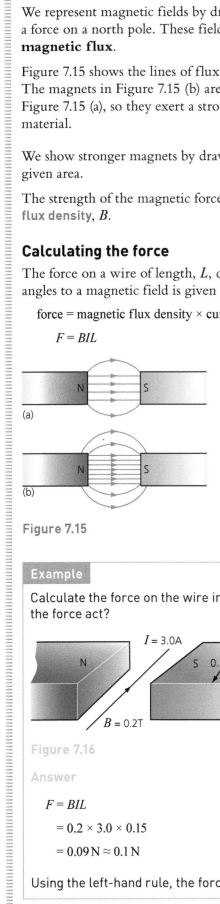

(a)

(b)

Figure 7.15

Example

Calculate the force on the wire in Figure 7.16. In which direction does the force act?

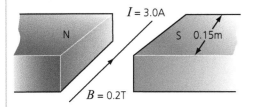

Figure 7.16

Answer

$$F = BIL$$
$$= 0.2 \times 3.0 \times 0.15$$
$$= 0.09\,\text{N} \approx 0.1\,\text{N}$$

Using the left-hand rule, the force acts downwards.

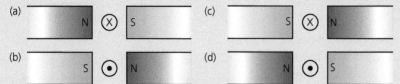

Now test yourself

13 Predict the direction of the force on the wire in each of the following cases.

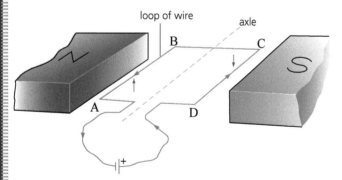

(a) [N ⊗ S]   (c) [S ⊗ N]

(b) [S ⊙ N]   (d) [N ⊙ S]

Figure 7.17

14 State two ways of increasing the force on a wire carrying a current in a magnetic field.
15 State the unit of magnetic flux density.
16 A wire of length 0.05 m is placed at right angles to a region of magnetic flux density 0.18 T. A current of 1.3 A flows through the wire.
   Calculate the force acting on the wire when
   (a) the field and current are at right angles to each other
   (b) the field and current are parallel.
17 In Figure 7.14, a magnetic force acts on the wire that carries a current. Newton's third law states that: whenever two objects interact, the forces they exert on each other are equal and opposite. On which object or objects does the wire exert a force?

Answers on p. 151

Electric motors

A coil carrying a current in a magnetic field tends to rotate. This is the basis of an electric motor.

In Figure 7.18 an upwards force acts on the wire AB, and a downwards force acts on the wire CD. These forces cause the coil to rotate clockwise. (Use the left-hand rule to check the directions of the force.) The coil stops rotating when it is vertical, because the two magnetic forces lie on the same vertical line.

Figure 7.18 **When the current flows the coil tends to rotate.**

Figure 7.19 shows the design of a motor that allows the coil to keep turning. A split-ring commutator allows the direction of the current in the coil to reverse, once the coil passes the vertical position. Now the forces act to keep the coil rotating in a clockwise direction.

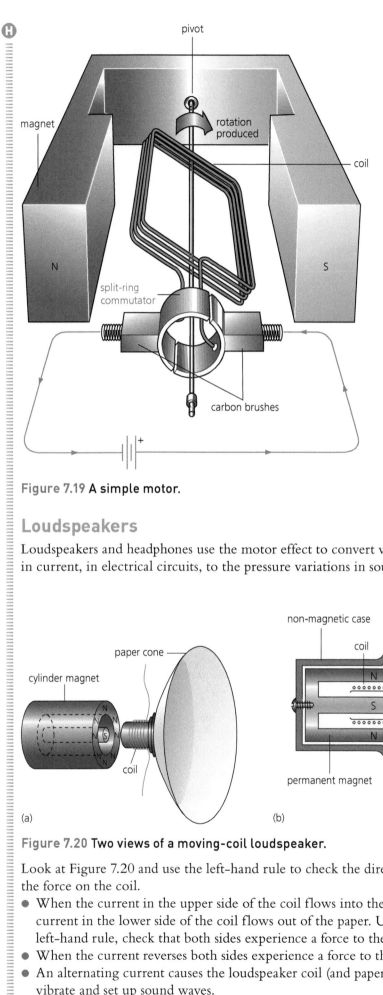

Figure 7.19 **A simple motor.**

A **split-ring commutator** allows the direction of current to reverse in a motor coil. This keeps the motor rotating in the same direction.

Loudspeakers

Loudspeakers and headphones use the motor effect to convert variations in current, in electrical circuits, to the pressure variations in sound waves.

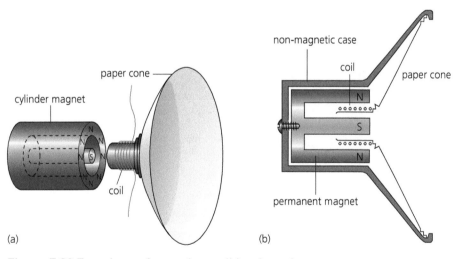

(a) (b)

Figure 7.20 **Two views of a moving-coil loudspeaker.**

Look at Figure 7.20 and use the left-hand rule to check the direction of the force on the coil.

- When the current in the upper side of the coil flows into the paper, the current in the lower side of the coil flows out of the paper. Using the left-hand rule, check that both sides experience a force to the left.
- When the current reverses both sides experience a force to the right.
- An alternating current causes the loudspeaker coil (and paper cone) to vibrate and set up sound waves.

18 Refer to Figure 7.18.
  (a) State the size of the force on the wire BC.
  (b) Explain why the coil will rotate and stop in a vertical position. You can draw diagrams to help your explanation.
19 Refer to Figure 7.19.
  (a) State three factors that affect the size of the forces that turn the coil.
  (b) Explain the function of the split-ring commutator.
20 (a) Make a sketch of a moving coil loudspeaker.
  (b) Use your sketch to explain why an alternating current causes the coil in the loudspeaker to vibrate.

Answers on pp. 151–152

Induced potential, transformers and the National Grid

REVISED

When a conducting wire moves through a magnetic field, a potential difference is induced across the ends of the wire; this is called an **induced potential difference**.

A potential difference is also induced if a magnetic field changes around a stationary conducting wire. If the wire is part of a complete circuit, the potential difference will cause a current in the circuit. This is an **induced current**.

Inducing a p.d. or a current in this way is called the **generator effect**, because this is how we generate electricity.

Investigating induced potential differences

Figure 7.21 shows how we can investigate the size and direction of induced potential differences and currents.

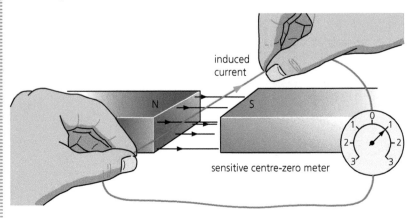

induced current

N S

sensitive centre-zero meter

Figure 7.21

First the wire must be moved up and down at right angles to the lines of magnetic flux. If the wire is moved parallel to the field lines, there is no current.

The induced current is increased when:
● the wire is moved faster
● stronger magnets are used.

The direction of the induced current is reversed when:
● the direction of movement is reversed
● the direction of the magnetic field is reversed.

⊕Coils and magnets

An induced current generates a magnetic field that opposes the original change: either the movement of the conductor or the change in magnetic field. This principle is demonstrated in Figure 7.22.

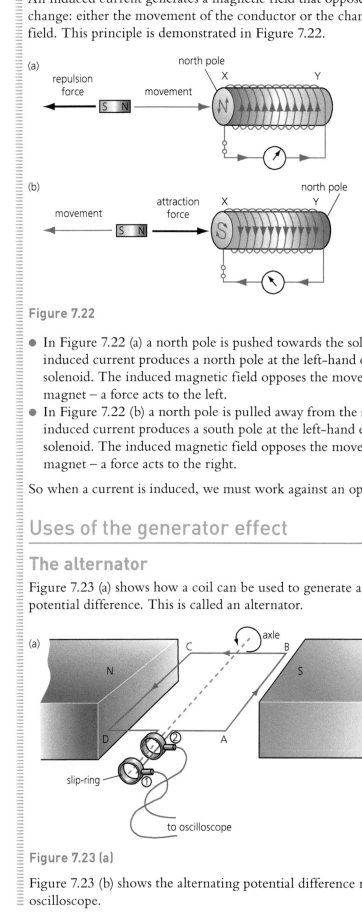

Figure 7.22

- In Figure 7.22 (a) a north pole is pushed towards the solenoid. The induced current produces a north pole at the left-hand end of the solenoid. The induced magnetic field opposes the movement of the magnet – a force acts to the left.
- In Figure 7.22 (b) a north pole is pulled away from the solenoid. The induced current produces a south pole at the left-hand end of the solenoid. The induced magnetic field opposes the movement of the magnet – a force acts to the right.

So when a current is induced, we must work against an opposing force.

Uses of the generator effect

REVISED ☐

The alternator

Figure 7.23 (a) shows how a coil can be used to generate an alternating potential difference. This is called an alternator.

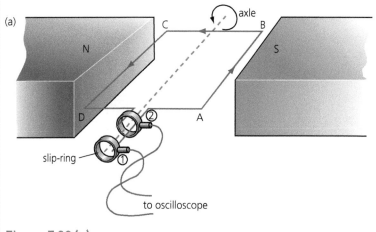

Figure 7.23 (a)

Figure 7.23 (b) shows the alternating potential difference recorded on an oscilloscope.

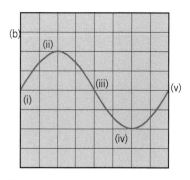

Figure 7.23 (b)

- The coil is vertical at times (i), (iii) and (v), and the induced potential difference is zero.
- The coil is horizontal at times (ii) and (iv). But the sides of the coil are moving in different directions at times (ii) and (iv), so the potential differences have different signs.

The size of the induced potential difference can be increased by:
- rotating the coil faster (see Figure 7.24)
- using stronger magnets
- putting more turns of wire in the coil
- wrapping the wire round a soft iron core.

The dynamo

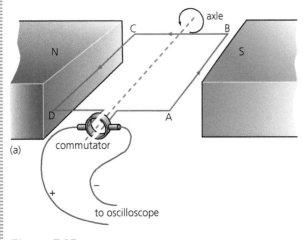

(a)

Figure 7.25

Figure 7.25 (a) shows the design of a dynamo that generates direct current. The design is similar to an alternator, but the split-ring commutator ensures that the current always flows the same way around the coil.

Figure 7.25 (b) shows the potential difference induced by the dynamo.

The microphone

Figure 7.26 shows how a moving coil microphone works.
- The sound waves reach the diaphragm and cause it to vibrate.
- The diaphragm is attached to a small coil that vibrates in a magnetic field.
- An alternating potential difference is induced across the coil. This can be amplified to drive a current through a loudspeaker, so we can hear the sound.

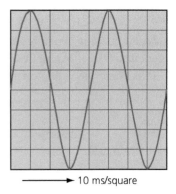

10 ms/square

Figure 7.24 The coil is now rotated twice as fast. The maximum potential difference and frequency of the potential difference are doubled.

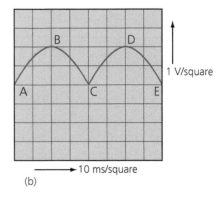

1 V/square

10 ms/square

(b)

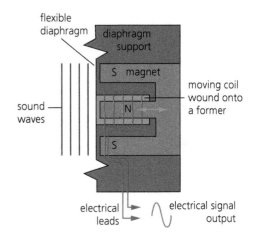

Figure 7.26

21 A wire is moved through a magnetic field so that a potential difference is induced across its ends.
   (a) State two ways in which the potential difference can be increased.
   (b) State the ways in which the direction of the potential difference can be reversed.

22 This question refers to the magnet and solenoid shown in Figure 7.27. The meter is a centre-zero meter.
   (a) State the size of the deflection on the meter when the magnet is stationary near the coil.
   (b) The magnet is now moved as follows. State the direction of the meter deflection in each case.
      (i) A north pole is moved away from B.
      (ii) A south pole is moved towards B.
      (iii) A north pole is moved towards A.
      (iv) A south pole is moved away from A.
   (c) State what two changes you could make to produce a larger deflection on the meter, using the same solenoid.

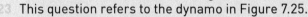

Figure 7.27 **The meter deflects to the left when a north pole moves towards end B.**

23 This question refers to the dynamo in Figure 7.25.
   (a) State whether the coil lies horizontal (as shown in the diagram) or vertical for each of the times A, B, C, D and E.
   (b) Copy the graph and add sketches to show the potential difference when:
      (i) the coil rotates in the opposite direction at the same speed
      (ii) the coil rotates in the same direction at twice the speed.

24 Explain how sound waves cause a microphone to produce an alternating potential difference.

Answers on p. 152

Transformers
REVISED

A transformer is made by putting two coils of wire onto an iron core, as shown in Figure 7.28. Iron is used for the transformer core because it is easily magnetised.

Transformers work using an alternating supply only. An alternating current (a.c.) supplied to the primary coil produces a changing magnetic field. The iron core links the changing magnetic field to the secondary coil, which induces an alternating potential difference in that coil. This is like moving a magnet in and out of a coil (see Figure 7.22).

If a direct current is supplied to the primary coil, the magnetic field is constant. A potential difference is not induced in the secondary coil. This is like leaving a magnet stationary in a coil.

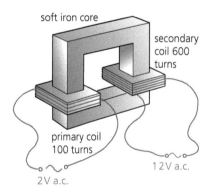

Figure 7.28

Step-up and step-down transformers

Transformers are useful because they allow us to change the potential difference of an a.c. supply.
- When there are more turns on the secondary coil than on the primary coil, the secondary p.d. is greater than the primary p.d. This is a **step-up** transformer.
- When there are fewer turns on the secondary coil than on the primary coil, the secondary p.d. is less than the primary p.d. This is a **step-down** transformer.

Step-up $V_s > V_p$

Step-down = $V_s < V_p$

(H) The rule for calculating the potential differences in a transformer is:

$$\frac{V_p}{V_s} = \frac{n_p}{n_s}$$

V_p is the p.d. across the primary coil

V_s is the p.d. across the secondary coil

n_p is the number of turns in the primary coil

n_s is the number of turns in the secondary coil

Example

A power station produces an alternating p.d. of 25 000 V. It is required to step up the p.d. to 400 000 V. A transformer with 3000 turns in its primary coil is to be used. Calculate the number of turns needed in the secondary coil.

Answer

$$\frac{V_p}{V_s} = \frac{n_p}{n_s}$$

$$\frac{25000}{400000} = \frac{3000}{n_s}$$

$$n_s = 3000 \times 16$$

$$= 48\,000 \text{ turns}$$

Power in transformers

Transformers transfer power from the primary coil to the secondary coil very efficiently. For a transformer that is 100% efficient:

power input (to the primary coil) = power output (from the secondary coil)

$$V_p I_p = V_s I_s$$

V_p is the primary p.d.

V_s is the secondary p.d.

I_p is the primary current

I_s is the secondary current

The National Grid

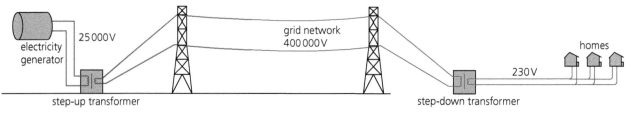

Figure 7.29

- When electricity is transmitted around the country it is stepped up to a very high p.d.
- When power is transmitted at a high p.d. the current flowing in the power lines is lower.
- With a low current in the power lines, much less power is wasted.

$$\text{power lost} = VI$$

$$= IR \times I$$

$$= I^2R$$

So when the current is low, much less energy is dissipated.

Now test yourself

25 (a) Explain what is meant by a *step-up transformer*.
   (b) State the relationship between the number of turns on the primary and secondary coils of a step-up transformer.
   (c) Give an example of where a step-down transformer is used in the home.

26 Fill in the gaps in the table below, where some information is given about four transformers.

Primary turns	Secondary turns	Primary p.d. in volts	Secondary p.d. in volts	Step-up or step-down
200	40		6	
400	10000	10		
	60	230	11.5	
24000		46000	230	

27 (a) Explain why transformers only work with an a.c. supply.
   (b) Explain:
      (i) why transformers are used to step up the p.d. from power stations to the power lines
      (ii) why transformers are used to step down the p.d. from the power lines for domestic use.

28 Use the information in Figure 7.30 to calculate the current flowing through the ammeter.

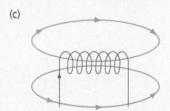

n_p = 400 turns n_s = 20 turns

Figure 7.30

Answers on p. 152

Summary

- Magnets have two poles, north and south.
- Like poles repel, unlike poles attract.
- Magnetic materials include: iron, steel, cobalt and nickel.
- A permanent magnet produces its own magnetic field. Two permanent magnets can attract or repel each other.
- An induced magnet becomes magnetic in a magnetic field. Induced magnets are always attracted towards permanent magnets.
- Magnetic fields: you should be able to sketch the shapes of the field close to a bar magnet, a wire carrying a current and a solenoid.

(c)

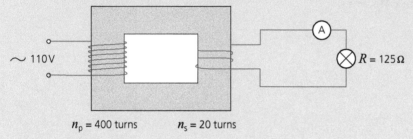

Figure 7.31 (a) The magnetic field around a bar magnet. (b) The magnetic field around a wire carrying current into the paper. (c) The magnetic field around a solenoid.

- The right-hand grip rule, Figure 7.9, shows you the direction of the magnetic field round a wire.
- An electromagnet can be made by putting an iron coil inside a solenoid.
- H ● Fleming's left-hand rule (Figure 7.14) shows the direction of the force on a current carrying wire, placed in a magnetic field:
   thuMb – Motion; First – Field; seCond – Current. This is the motor effect.

(a)   (b)

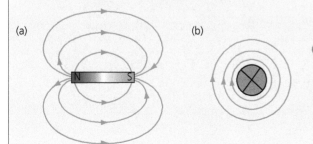

- The force on a wire, length L, carrying current, I, at right angles to a magnetic field of flux density, B, is:

$$F = BIL$$

- The motor effect (as described by Fleming's left-hand rule) is the principle behind the electric motor – a coil in a magnetic field tends to turn.
- The split-ring commutator allows a coil to turn continuously.
- An alternating current causes a loudspeaker coil to vibrate (Figure 7.20).
- When a conducting wire moves through a magnetic field (or the magnetic field changes), a potential difference is induced across its ends.
- An induced current produces a magnetic field, which opposes the motion that induced the current (Figure 7.22).
- The alternator generates alternating current (Figure 7.23). The faster the alternator rotates, the larger and induced potential difference, and the higher the frequency.

- A dynamo generates direct current.
- Sound waves generate alternating potential differences in a microphone (Figure 7.26).
- Transformers step up or step down alternating potential differences.
- For transformers:

$$\frac{V_p}{V_s} = \frac{n_p}{n_s}$$

and

$$V_p I_p = V_s I_s$$

- Transformers are used in the National Grid to step potential differences up to very high values (400000V), then power is transmitted with low currents and less energy is dissipated.

Exam practice

1 (a) State the unit of magnetic flux density. [1]
  (b) Iron and steel are two types of magnetic material.
    (i) Which material can be used as an induced magnet? [1]
    (ii) Which material can be used as a permanent magnet? [1]
  (c) (i) Sketch the shape of the magnetic field around a bar magnet. [2]
    (ii) Explain how you would use a compass to plot the shape of the magnetic field. [2]
  (d) Figure 7.32 shows two electromagnets A and B, which are suspended on threads and are free to move.

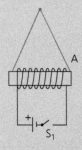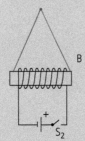

Figure 7.32

Figure 7.33

  (i) What will happen when switches S_1 and S_2 are closed? Will the magnets attract or repel? [1]
  (ii) What will happen when one of the batteries is reversed, and the switches are closed? [1]

2 (a) Copy Figure 7.33 and sketch the shape of the magnetic field around the wire. [2]
  (b) The flux density between the poles of the magnet in Figure 7.34 (a) is 0.1 T.
    (i) Use the information in Figure 7.34 (a) to calculate the size of the force on the wire. [2]
    (ii) Determine the direction of the force on the wire. [1]

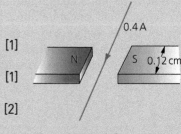

Figure 7.34 (a)

→

(c) Explain what will happen to the coil of wire shown in Figure 7.34 (b), when the current is switched on. [3]

(d) (i) A split-ring commutator is added to the coil of wire as shown in Figure 7.34 (c). Explain how the split-ring commutator allows the coil to turn continuously. [2]

   (ii) State two changes to the motor design that would make the motor turn faster. [2]

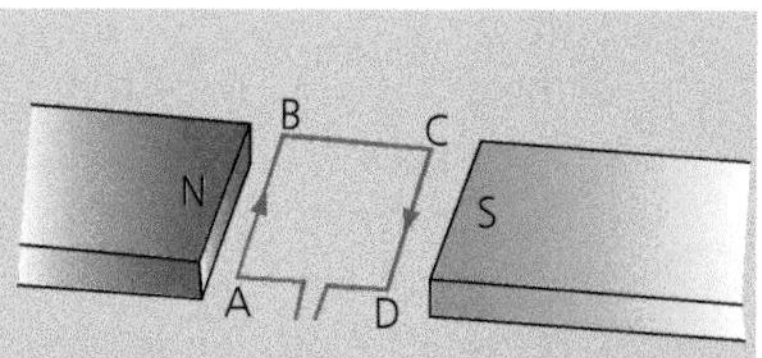

Figure 7.34 (b)

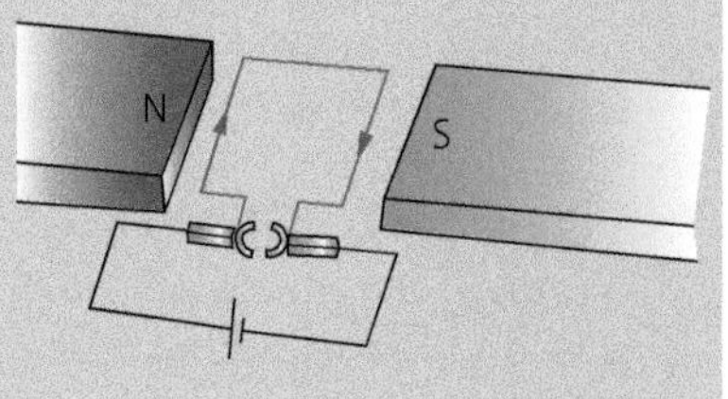

3 When the magnet is pushed towards the coil as shown in Figure 7.35 (a), the meter deflects towards the left.

(a) Describe what happens when:

   (i) the magnet is stationary inside the coil [1]

   (ii) the south pole is pulled away from the coil [1]

   (iii) the north pole is pushed towards the coil. [1]

(b) Figure 7.35 (b) shows a generator that is attached to the blades of a wind turbine.

   (i) Explain why the rotating magnet generates an a.c. potential difference. [2]

   (ii) Explain why the coils mounted on a soft iron core. [1]

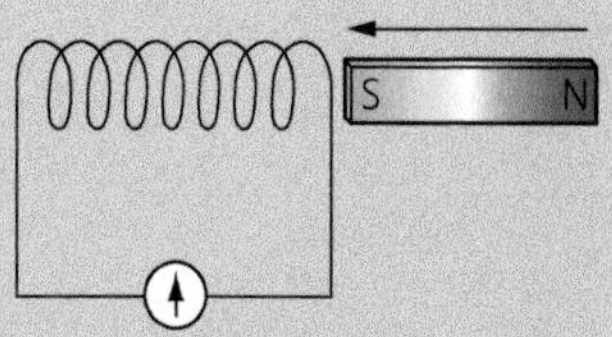

Figure 7.34 (c)

Figure 7.35 (a)

Figure 7.35 (c)

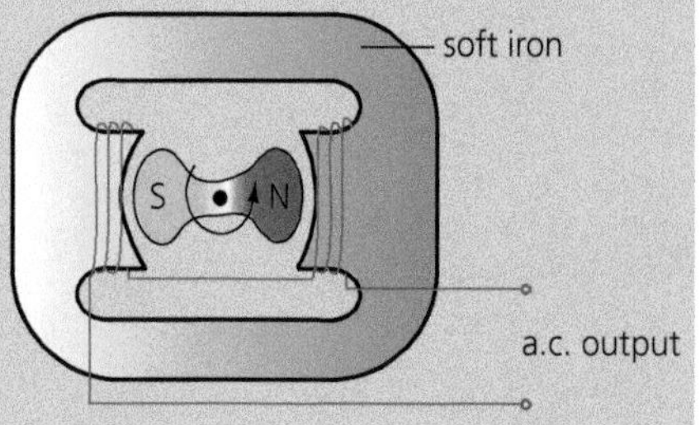

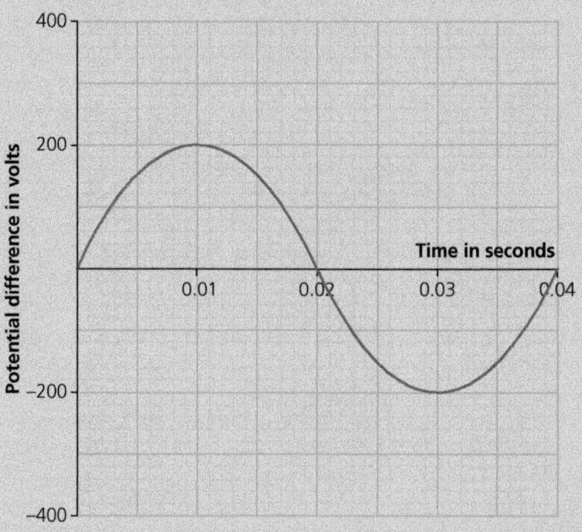

Figure 7.35 (b)

Figure 7.35 (c) shows the potential difference when the wind is blowing.

   (iii) State the time period of one cycle of the potential difference. [1]

   (iv) Use your answer to part (b) (iii) to calculate the frequency of the a.c. output. [1]

   (v) The wind blows faster on another day and the generator rotates twice as quickly. Copy Figure 7.35 (c) and show the induced potential difference now. [2]

4 In Figure 7.36 a heavy copper pendulum suspended from a wire swings backwards and forwards from A to C, between the poles of a magnet. As the wire moves an induced current flows through the resistance R, and the data logger records the potential difference across R.

(a) Give two reasons why the induced current is greatest as the pendulum moves past point B. [2]

(b) Sketch a graph to show how the induced potential difference changes as the pendulum goes from A to B to C and back again. [3]

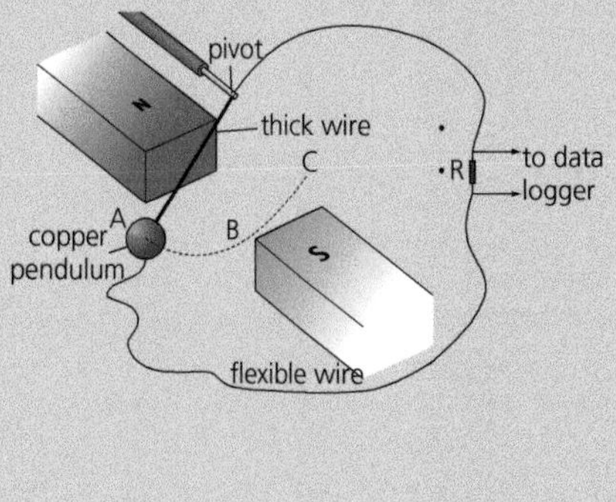

Figure 7.36

5 The diagram shows a step-down transformer in the plug of an electric shaver, which is used in a mains bathroom socket.

(a) Use the information in the diagram to show the output potential difference is about 12V. [3]

(b) When the shaver is working normally, the current through it is 0.45A. Calculate the current flowing into the transformer from the mains. [3]

(c) Explain why transformers only work using an alternating current supply. [3]

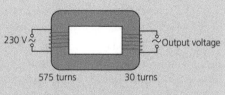

230 V   Output voltage

575 turns   30 turns

Figure 7.37

6 Transformers are used to step up potential differences, generated at power stations, to about 400 000 V for transmission over long distances. Explain why. [4]

7 Figure 7.38 (a) shows a device called a current balance. A wire is balanced on supports so that the weight of the insulator and wire on the left-hand side of the supports is balanced by the weight of the wire on the right-hand side.

When the switch is closed, there is a force downwards on the wire AB.

A small weight of 7.2×10^{-2} N is placed on the insulating bar to return the wire AB to its original position, as shown in Figure 7.38 (b). The current flowing when the wire AB is back in its original position is 2.0A. Use the information in the two diagrams to calculate the magnetic flux density between the poles of the magnets. Give the correct unit for magnetic flux density. [6]

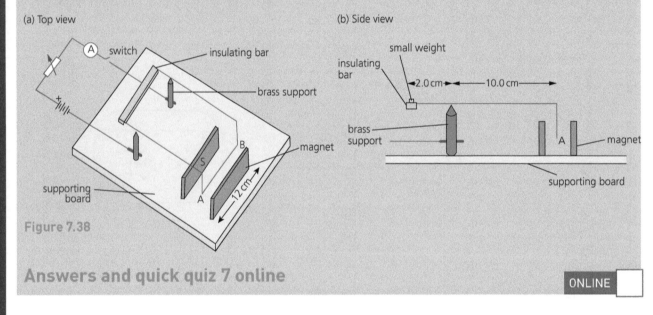

(a) Top view

switch
insulating bar
brass support
magnet
supporting board

(b) Side view

small weight
insulating bar
2.0 cm   10.0 cm
brass support
A
magnet
supporting board

Figure 7.38

Answers and quick quiz 7 online

ONLINE

8 Space physics

Our Solar System

The Sun is a star which is at the centre of the Solar System. The Sun releases energy by the process of nuclear fusion.

The Earth is one of eight planets that are in stable orbits around the Sun. There are also five dwarf planets that orbit the Sun.

Some of the planets also have natural satellites, or moons, in orbit around them.

Our Moon is a natural satellite of the Earth.

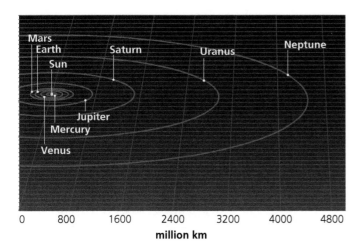

Figure 8.1 **The orbits of the planets around the Sun.**

Galaxies

Our Solar System is a very small part of the Milky Way galaxy. This galaxy contains about 200 000 million stars. Many of these stars are thought to have their own system of planets.

The Universe contains many billions of galaxies.

Now test yourself

TESTED

1. Explain the meaning of the term *Solar System*.
2. Explain the difference between a dwarf planet and a moon.
3. What is the name of our galaxy?

Answers on p. 152

The life cycle of a star

REVISED

Our Sun, together with the planets and moons, was formed about 4.6 billion years ago.

Birth of a star

- A star like our Sun is formed from a large cold cloud of dust and gas. This is called a nebula.

- The pull of gravity acts on the cloud and causes it to collapse. As the cloud collapses the atoms and molecules move very quickly.
- When molecules collide with each other, their kinetic energy is transferred to the thermal store of the gas. The gas reaches a temperature of several million degrees Celsius.
- The collapsing and heating ball of gas is called a protostar.
- When the temperature is high enough, hydrogen nuclei (protons) are able to collide and fusion begins. A star has been born.

The stable period of a star

The fusion reactions inside stars lead to an equilibrium. The pull of gravity tends to collapse the star, but this is balanced by the outward pressure due to the nuclear fusion.

A star the size of the Sun will remain in a stable state for billions of years. A stable star producing energy by fusing hydrogen nuclei is called a **main sequence star**.

The death of a star

The way a star ends its life depends on its size.

Stars about the same size as the Sun

- Towards the end of a star's life, the supply of hydrogen begins to run out. Without the pressure of fusion, the star starts to collapse.
- The further collapse of the star makes the core of the star even hotter – up to 100 million °C. Now helium can fuse to form heavier elements such as carbon and oxygen.
- The hot core causes the outer surface of the star to swell into a **red giant**, which can have a radius 100 times that of the Sun and can be 1000 times brighter than the Sun.
- Eventually, fusion reactions can no longer happen. Now the star collapses to form a **white dwarf**, which is not much larger than the Earth.

The star's life is over. It cools down and eventually becomes a cold **black dwarf**.

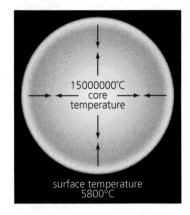

Figure 8.2 The outward forces from the nuclear fusion balance the gravitational forces tending to collapse the star.

A **main sequence star** is one that releases energy by fusing hydrogen into helium.

A **red giant** is a very large star which fuses helium into heavier elements.

A **white dwarf** is a star at the end of its life. No fusion occurs, it is cooling down.

A **black dwarf** is a white dwarf that has cooled down.

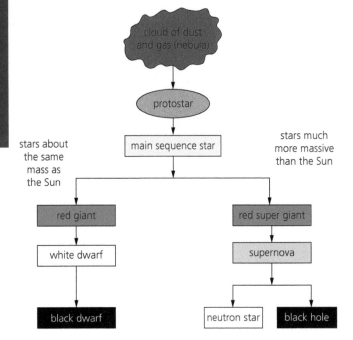

Figure 8.3

Stars much larger than the Sun

- When a very large star reaches the end of its main sequence stage, it collapses in the same way as a smaller star, but then expands to form a **red super giant**. Such a star can have a radius 1000 times larger than the Sun and can be 100 000 times brighter.
- The red super giant is able to fuse nuclei to form elements as massive as iron.
- When the red super giant runs out of its nuclear fuel it collapses very quickly. The rapid collapse creates such high temperatures that the star explodes like a cosmic nuclear bomb. This is a **supernova**. The energy is so great that elements with an atomic number higher than iron are formed. Our Solar System was formed out of the remnants of a supernova – this is why we have heavy elements such as gold and uranium.
- At the same time as the supernova explosion, great gravitational forces collapse the centre of the star. Either the core is left as: (i) a **neutron star**, which has a diameter of a few kilometres – this is made entirely of neutrons; (ii) the core becomes a **black hole** – this is a microscopic point, so dense that not even light can escape.

> A **red super giant** is a large red giant, which has a very hot core that can produce elements as heavy as iron by fusion.
>
> A **supernova** is a gigantic explosion caused by runaway fusion reactions.
>
> A **neutron star** is a very small dense star made out of neutrons.
>
> A **black hole** is the most concentrated form of matter from which not even light can escape.

Now test yourself

TESTED

4 The various stages of the life cycle of a star much larger than the Sun are listed below. Put them in the correct order.
   A Neutron star
   B Nebula
   C Red super giant
   D Supernova
   E Protostar
   F Main sequence star
5 (a) Name the process that allows the Sun to emit its own light.
   (b) (i) At what stage of its life is the Sun now?
       (ii) What will be the final stage of the Sun's life?
6 (a) Explain what is meant by the term *main sequence star*.
   (b) A main sequence star is in a state of equilibrium. Explain what this means and illustrate your answer with a diagram.
7 Write an account of the life cycle of a star which is about the same mass as our Sun.

Answers on p. 152

Orbital motion, natural and artificial satellites

REVISED

Moons and planets move in nearly circular orbits.
- Gravity causes planets to orbit the Sun.
- Gravity causes the Moon and artificial satellites to orbit the Earth.

Figure 8.4 shows the pull of gravity from the Earth, which keeps the Moon in orbit. In position M_1, the Moon has a speed v; the pull of gravity deflects its motion. Later the Moon has moved to position M_2, but it still has the same speed, v. The force of gravity does not make the Moon travel any faster, but the force changes the direction.

Orbital motion and velocity

- In Figure 8.4 the speed v of the Moon remains constant.
- But because the direction of motion changes there is a change of velocity – remember velocity is a vector quantity.

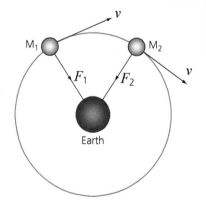

Figure 8.4 The force of gravity on the Moon, towards the Earth, keeps the Moon in its orbit.

- Because there is a velocity change, the Moon is accelerating all the time. But this acceleration does not speed the Moon up, it causes the change of direction.
- The gravitational force causes the acceleration.

Speed of orbit

Figure 8.5 shows the importance of speed. S is a satellite travelling round the Earth.

- If the speed is too great, the satellite follows path C and disappears into outer space.
- If the speed is too slow, the satellite follows path A and falls to Earth.
- When the satellite moves at the right speed, it follows path B and is in a stable orbit.

The speed of a moon, planet or satellite can be calculated if the distance moved in one orbit is known, using the equation:

$$\text{speed} = \frac{\text{distance moved in one orbit}}{\text{time}}$$

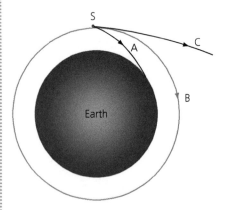

Figure 8.5

Exam tip

$$\text{acceleration} = \frac{\text{change of velocity}}{\text{time}}$$

Velocity is a speed in a given direction.
So when the Moon changes direction, it accelerates because its velocity changes.

Typical mistake

Do not say the velocity of a planet is constant in its orbit. The speed is constant, the velocity changes because the direction of motion changes.

Now test yourself

TESTED

8 A satellite is an orbit around the Earth.
   (a) Name the force that keeps the satellite in orbit.
   (b) Describe the orbit of the satellite.
9 A satellite is in an orbit around the Earth. The distance travelled in one orbit is 50 000 km and it takes 2 hours to complete one orbit. Calculate the speed of the orbit in m/s.
10 (a) Explain the difference between speed and velocity.
   (b) Explain why the velocity of a moon changes as it orbits a planet.

Answers on pp. 152–153

Red-shift

We know from experience that the sound we hear from the siren of a police car changes pitch as the car passes us. When the car moves away from us, we hear a lower frequency and the wavelength of the waves increases.

Figure 8.6 A teacher whirls a small loudspeaker round his head to model the effect of the red-shift. When the loudspeaker moves away from the students they hear a decrease in pitch – which means an increase in wavelength.

When a light source is moving away from us very quickly, the wavelength of the light also increases – it moves towards the red end of the spectrum. This is called a red-shift. The light that we see from distant galaxies is red-shifted. This tells us that these galaxies are moving away from us.

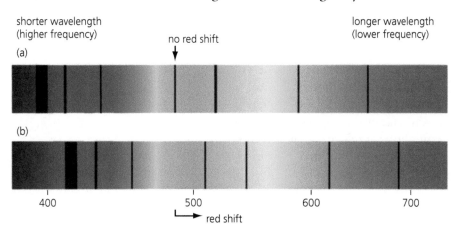

Figure 8.7 (a) This image shows light which is emitted from the Sun. It is crossed by black lines. (b) This second image shows light emitted by a distant galaxy. It has the same pattern of lines. But the pattern of lines has been shifted towards the red end of the spectrum. This red-shift tells us the galaxy is moving away from us.

The Big Bang Theory

Edwin Hubble was the first astronomer to look at the red-shift. He discovered the following facts.

- Light from distant galaxies is red-shifted, so they are moving away from us.
- Galaxies in all directions are moving away from us.
- The further a galaxy is away from us, the greater its red-shift. So the more distant the galaxy, the greater its speed.

This idea led to the Big Bang Theory: the Universe began as a very small region (or point) that was extremely hot and dense. This region exploded throwing matter outwards. The matter that moved the fastest has travelled the furthest away from the original explosion.

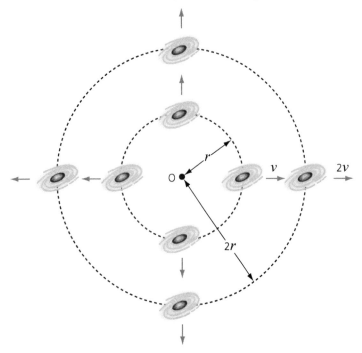

Figure 8.8 The Solar System is placed at O. Galaxies are moving away from us in all directions. The more distant galaxies are moving faster. This suggests that at one instant, a long time in the past, all the galaxies were in the same position.

The Big Bang Theory is accepted today as the best theory for the origin of the Universe. However, not everything is understood about the Universe. For example, recent observations of supernova explosions suggest that distant galaxies are receding faster than expected. It appears that the expansion of the Universe might be accelerating. We would expect the expansion to slow down due to the pull of gravity. These observations have led to theories about dark energy that might be pushing the Universe outwards.

Now test yourself

TESTED ☐

11 An astronomer discovers that light from a distant galaxy has been shifted towards the red end of the spectrum. Explain what she can deduce from this discovery.

12 Which of the following statements about the Big Bang Theory are true and which are false?
  A It has been proved correct by scientists.
  B It is supported by the fact that most galaxies are moving away from the Earth.
  C It is supported by recent evidence looking at supernovas.
  D It is the only way to explain the origin of the Universe.
  E Most scientists think it is the best way to explain the origin of the Universe.

13 Summarise the evidence for the Big Bang Theory for the origin of the Universe.

Answers on p. 153

Summary

- The Solar System is a small part of our local galaxy – the Milky Way.
- Eight planets (including the Earth) and five dwarf planets orbit the Sun.
- Moons orbit planets.
- Stars have different life cycles according to their mass.
- A star the mass of the Sun follows this cycle: gas nebula → protostar → main sequence star → red giant → white dwarf → black dwarf.
- A star much more massive than the Sun follows this cycle: gas nebula → protostar → main sequence star → super red giant → supernova → neutron star or black hole

- The force of gravity keeps planets, satellites and moons in orbit.
- In a circular orbit planets, moons and satellites move at constant speeds, but their velocity changes because they change direction.
- Galaxies moving quickly away from us show an increase in the wavelength of light. This is called the red-shift.
- The further away the galaxies, the faster they are moving and the greater the increase of wavelength (and red-shift).
- The observed red-shift provides evidence that the Universe is expanding and supports the Big Bang Theory.

Exam practice

1 (a) Explain the meaning of the term *Solar System*. [1]
  (b) Name three types of bodies to be found in our Solar System. [3]
  (c) What is the name of our galaxy? [1]
2 Explain the following terms:
  (a) *main sequence star* [2]
  (b) *super red giant*. [2]
3 (a) Name the process by which a star produces energy so that it can emit light and other electromagnetic waves. [1]
  (b) Explain why a main sequence star remains stable for a long time. [2]
4 Give an account of the life cycle of a star which is much more massive than the Sun. [6]
5 Figure 8.9 shows two of Jupiter's moons, Io and Europa, in orbit around Jupiter.

  (a) Explain the difference between a moon and a planet. [1]
  The table below gives some information about the orbits of the moons.

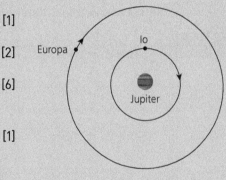

Figure 8.9

	Distance travelled in one orbit (km)	Time period of orbit (h)
Io	2 600 000	42.5
Europa	4 200 000	85.0

  (b) Use the data in the table to show that Io travels at a greater speed than Europa. (Express your answers in km/h.) [3]
  (c) Explain what would happen to Io if it travelled at the same speed as Europa. [2]
  (d) A student makes this statement. Explain what is wrong with it. [2]

  'Io and Europa travel with constant velocities around Jupiter.'

→

6 In 1929 the astronomer Edwin Hubble measured the speeds of galaxies and their distances away from the Earth.
Figure 8.10 shows some of his results.

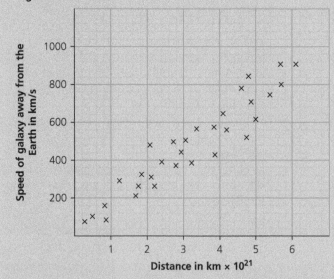

Figure 8.10

(a) What relationship between distance and speed is suggested by this data? [1]
(b) Galaxy A is at a distance of 2×10^{21} km; Galaxy B is at a distance of 4×10^{21} km from us.
Explain which galaxy shows the greater red-shift in its light. [2]
(c) Hubble's data supports a theory for the origin of the Universe, which suggests the Universe originated from a small point.
(i) What is the name of this theory? [1]
(ii) Explain why Hubble's data supports this theory. [4]

Answers and quick quiz 8 online

ONLINE

Now test yourself answers

Chapter 1 Energy

1 (a) Thermal store of water
   (b) Chemical store of the battery
   (c) Chemical store of the petrol
   (d) Kinetic store of the train
   (e) Kinetic and gravitational potential stores of the ball
   (f) Elastic store of the spring

2 (a) Kinetic energy stored in the car is transferred to thermal energy stored in the brakes and the surroundings.
   (b) Chemical energy stored in the fuel is transferred to kinetic energy and gravitational potential energy stored in the plane. As the fuel burns some energy is also transferred to the thermal store of the surroundings.
   (c) Chemical energy stored in the cell is transferred to the thermal energy store of the surroundings.
   (d) Thermal energy stored in the coffee is transferred to the thermal store of the surroundings.

3 B   60 J (E_p)
   C   60 J (E_k)
   D   90 J (E_k)

4 $E_p = mgh$
   $= 45 \times 9.8 \times 310$
   $= 140 \, kJ$

5 $E_k = \frac{1}{2} mv^2$
   $= \frac{1}{2} \times 0.02 \times (700)^2$
   $= 4900 \, J$

6 Change in
   $E_k = \frac{1}{2} mv_2^2 - \frac{1}{2} mv_1^2$
   $= \frac{1}{2} \times 900 \times 18^2 - \frac{1}{2} \times 900 \times 12^2$

$= 81\,000 \, J$
$= 81 \, kJ$

7 $E_e = \frac{1}{2} ke^2$
   $= \frac{1}{2} \times 1800 \times (0.1)^2$
   $= 9.0 \, J$

8 (a) $E_p = mgh$
      $= 0.08 \times 9.8 \times 12$
      $= 9.4 \, J$
   (b) $E_k = 9.4 \, J$
      $9.4 = \frac{1}{2} mv^2$
      $9.4 = \frac{1}{2} \times 0.08 v^2$
      $v^2 = \frac{9.4}{0.04}$
      $v = 15 \, m/s$
   (c) The gravitational potential energy has been transferred into the thermal store of the river.

9 (a) $E_p = mgh$
      $= 52 \times 9.8 \times 4$
      $= 2038 \, J \approx 2.0 \, kJ$
   (b) $2038 \, J \approx 2.0 \, kJ$
   (c) $E_e = \frac{1}{2} ke^2$
      $2038 = \frac{1}{2} k \times (0.9)^2$
      $k = \frac{2 \times 2038}{(0.9)^2}$
      $= 5030 \, N/m \approx 5000 \, N$

10 (a) $E_e = \frac{1}{2} ke^2$
      $= \frac{1}{2} \times 300 \times (0.9)^2$
      $= 121.5 \, J \approx 120 \, J$
   (b) $E_k = \frac{1}{2} mv^2$
      $121.5 = \frac{1}{2} \times 0.03 v^2$
      $v^2 = \frac{121.5 \times 2}{0.03}$
      $v = 90 \, m/s$

11 (a) newton, N
   (b) watt, W
   (c) joule, J
   (d) joule, J

12 (a) $E_p = mgh$

$= 110 \times 9.8 \times 2.3$

$= 2480\,J \approx 2500\,J$

(b) power $= \dfrac{\text{energy transferred}}{\text{time}}$

$= \dfrac{2480}{1.7}$

$= 1460\,W \approx 1500\,W$

13 (a) $W = mg$

$= 120 \times 3.7$

$= 444\,N \approx 440\,N$

(b) $E_p = mgh$

$= 444 \times 8$

$= 3552\,J \approx 3600\,J$

(c) Chemical energy stored in his muscles is transferred to gravitational potential energy and thermal energy stored in his body.

14 $P = \dfrac{\text{work done}}{\text{time}}$

$= \dfrac{15\,000 \times 28}{84}$

$= 5000\,W = 5.0\,kW$

15 power $= \dfrac{E}{t}$

$= \dfrac{3000}{120}$

$= 25\,W$

16 J/kg°C

17 (a) $\Delta E = mc\Delta\theta$

$= 90 \times 1000 \times 16$

$= 1\,440\,000\,J = 1.4\,MJ$

(b) (i) work = power × time

$= 2200 \times 120$

$= 264\,000\,J \approx 260\,kJ$

(ii) 260 kJ. This assumes that no thermal energy escapes to the surroundings.

(iii) $\Delta E = mc\Delta\theta$

$264\,000 = 0.8 \times 4200 \times \Delta\theta$

$\Delta\theta = \dfrac{264\,000}{0.8 \times 4200}$

$= 79°C$

So the final temperature is: 79 + 18 = 97°C

(c) $\Delta E = mc\Delta\theta$

$12\,000 = 1 \times c \times 30$

$c = 400\,J/kg°C$

18 (a) $\Delta E = mc\Delta\theta$

$= 0.2 \times 4000 \times 20$

$= 16\,000\,J$

$\Delta E = \text{power} \times \text{time}$

$16\,000 = 800 \times \text{time}$

time $= \dfrac{16\,000}{800}$

$= 20\,s$

(b) You have to warm the cup up as well, and that takes some of the thermal energy.

19 Dissipation means 'spread out' or 'wasted'.

20 (a) Loft insulation, cavity wall insulation, double glazing, carpets or just making sure doors and windows do not let cold air in.

(b) Air is a good insulator. All of the first four trap air; so heat is not conducted through the walls, windows, roof or floor.

21 Engines are made efficient so that they convert less energy to thermal energy.

Cars are streamlined to reduce drag.

Joints are lubricated to reduce friction.

22 (a) (i) Make sure we always use 200 ml of water.

(ii) Make sure the initial temperature is always the same.

(iii) Make sure the beaker is always on the same mat.

(iv) Keep the time the same for each test.

(b) (i) Keep the fair tests above, and then keep a lid on the beaker for each of the insulator tests.

(ii) The best insulator will show the lowest temperature change in the measured time.

(c) A possible source of random error is the reading taken of the temperature.

23 efficiency $= \dfrac{\text{useful energy output}}{\text{total energy input}}$

24 (a) Available energy $= 0.3 \times 4.5 \times 10\,MJ$

$= 13.5\,MJ \approx 14\,MJ$

(b) Energy dissipated = 45 MJ – 13.5 MJ

$= 31.5\,MJ \approx 32\,MJ$

25 (a) $E_p = mgh$

$= 60 \times 9.8 \times 35$

$= 20\,580\,J \approx 21\,kJ$

(b) E_p is only 25% of the energy transferred, 75% is transferred to thermal energy.

So the total energy transferred from the chemical store is:

$4 \times 20\,580\,J = 82\,320\,J \approx 82\,kJ$

(c) Dissipated energy $= \dfrac{3}{4} \times 82\,320\,J$

$= 61\,740\,J \approx 62\,kJ$

(d) $\Delta E = mc\Delta\theta$

$61\,740 = 60 \times 4000 \times \Delta\theta$

$\Delta\theta = \dfrac{61\,740}{60 \times 4000}$

$= 0.26°C$

26 (a) A renewable resource never runs out because it can be replenished. Examples: wind power, tidal power.

(b) A non-renewable resource will run out. Examples: oil, coal, gas.

27 (a) Coal is reliable and relatively cheap.

(b) Coal produces CO_2 which is a greenhouse gas. Other gases contribute towards pollution, e.g. SO_2 produces acid rain.

Chapter 2 Electricity

1

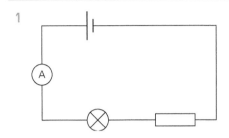

2 (a) $Q = It$

$12 = I \times 120$

$I = \dfrac{12}{120}$

$= 0.01\,A$

(b) $I = 100\,mA$

3 (a) For an ohmic resistor (at constant temperature) the current is proportional to the potential difference.

(b) The current is not proportional to the potential difference.

4 (a) $V = 0.85\,V$

(b) $V = IR$

$0.85 = 0.1 \times R$

$R = \dfrac{0.85}{0.1}$

$= 8.5\,\Omega$

5 (a) ohm

(b) coulomb

6 Set up the circuit as shown.

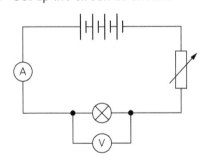

Take pairs of readings of potential difference across the lamp and the current through it. The potential difference needs to vary from 0 V up to about 6 V (for a 6 V rated lamp).

Then a graph of current against potential difference is plotted.

7 (a) $80\,\Omega$

(b) $2300\,\Omega$ or $2.3\,k\Omega$

8 Less than $20\,\Omega$

9 (a) $3\,V$

(b) $A_1 = 0.5\,A$; $A_2 = 1.2\,A$

10 (a) (i) $V = IR$

$= 40 \times 10^{-3} \times 250$

$= 10\,V$

(ii) p.d. across thermistor $= 12\,V - 10\,V = 2\,V$

$V = IR$

$2 = 40 \times 10^{-3} \times R$

$R = \dfrac{2}{40 \times 10^{-3}}$

$= 50\,\Omega$

(b) The temperature goes up, so the resistance of the thermistor goes down. Therefore the current in the circuit rises and the potential difference across the $250\,\Omega$ resistor rises: the voltmeter reading goes up.

11 (a) $A_1 = 0$; $A_2 = 0.2\,A$

(b) $V = IR$

$6 = 0.2 \times R$

$R = \dfrac{6}{0.2}$

$= 30\,\Omega$

(c) (i) $A_2 = 0.1 + 0.2 = 0.3\,A$

(ii) $6\,V$

(d) More current flows. Since $R = \dfrac{V}{I}$, a larger current means R is less.

(e) When the light intensity increases the resistance of the LDR goes down. So A_1 rises; A_3 is unchanged; but $A_2 = A_3 + A_1$ so A_2 rises.

12 (a) watt

(b) joule

(c) coulomb

13 (a) joule/coulomb

(b) coulomb/sec

14 (a) $P = VI$

$= 230 \times 8$

$= 1840\,W \approx 1800\,W$

(b) $P = VI$

$= 12 \times 5$

$= 60\,W$

15 $E = VIt$

$= 230 \times 15 \times 180$ (time must be in seconds)

$= 6.2 \times 10^5\,J$

16 (a) $E = VQ$

$= 12 \times 150$

$= 1800\,J$

(b) $E = VIt$

$= 6 \times 0.008 \times 4 \times 3600$

$= 691\,J \approx 690\,J$

17 $E = Pt$

$\quad = I^2Rt$

$\quad = (0.1)^2 \times 220 \times (20 \times 60)$

$\quad = 2640\,J \approx 2600\,J$

18 (a) Electrons have been transferred from the duster to the plastic rod.

(b) The duster is positively charged.

19 An electric field is a region in which a charged objected experiences a force. The force is a non-contact force. The field direction is the direction of a force on a positive charge in the field.

You can draw a diagram similar to Figure 2.26.

Chapter 3 Particle model of matter

1 A 400 kg/m³

B 0.8 m³

C 90 kg

D 2 × 10⁴ kg/m³

2 $V = 0.057 \times 0.057 \times 0.057\,m^3$

$\rho = \dfrac{M}{V}$

$\quad = \dfrac{144.6 \times 10^{-3}}{(5.7 \times 10^{-2})^3}$

$\quad = 781\,kg/m^3 \approx 780\,kg/m^3$

3 I would use a measuring flask to measure the volume of the object.

Step 1: Fill the flask to (say) 30 ml.

Step 2: Put the object in the flask, new volume (say) 50 ml.

Step 3: Calculate the volume = 20 ml

$density = \dfrac{mass}{volume}$

So measure the mass on an electronic balance and calculate the density. There are random errors which might occur in the measurement of the mass and the volumes of the water. The larger error will be in the measurement of the volumes of water. A more accurate result will be obtained by repeating the measurements of the volume of the object, and then calculating a mean value.

4 (a) Refer to Figure 3.1.

(b) The atoms in a gas are far apart, so there is a small mass in a large volume. In a solid there is a large mass in a small volume, so the density is greater:

$density = \dfrac{mass}{volume}$.

5 (a) Internal energy is the sum of the kinetic energies and potential energies of the atoms or molecules.

(b) The atoms can be made to move faster (in a random way).

The potential energies can be increased by separating the atoms – this happens when the substance melts or boils.

6 (a) Change of state occurs when a substance goes from one state to another, e.g. solid to a liquid.

(b) Ice melts; water boils.

7 (a) J/kg °C

(b) The specific heat capacity of a substance is the energy required to warm 1 kg of the substance by 1°C.

8 (a) $\Delta E = mc\Delta\theta$

$\quad = 0.5 \times 4200 \times 50$

$\quad = 1.05 \times 10^5\,J \approx 1.0 \times 10^5\,J$

(b) $\quad\quad \Delta E = mc\Delta\theta$

$\quad 1.26 \times 10^4 = 0.2 \times 630 \times \Delta\theta$

$\quad\quad \Delta\theta = \dfrac{1.26 \times 10^4}{0.2 \times 630}$

$\quad\quad\quad = 100°C$

(c) $\Delta E = mc\Delta\theta$

$\quad = 75 \times 1000 \times 25$

$\quad = 1.875 \times 10^6\,J \approx 1.9 \times 10^6\,J$

9 (a) $\quad\quad \Delta E = mc\Delta\theta$

$\quad 13.5 \times 10^3 = 0.8 \times c \times 34$

$\quad\quad c = \dfrac{13\,500}{0.8 \times 34}$

$\quad\quad\quad = 496\,J/kg\,°C \approx 500\,J/kg\,°C$

(b) Energy is usually dissipated to the surroundings. Energy is also required to warm the beaker and the thermometer.

10 (a) This is the energy required to melt one kilogram of the substance (at its melting point), without a change of temperature.

(b) J/kg

11 (a) Energy from our body evaporates the sweat. This cools us down.

(b) Panting evaporates saliva from the dog's tongue, thus removing thermal energy from the body.

(c) For the water to freeze, a large amount of energy must be extracted from it. Water has a high latent heat of fusion. So that store of energy is transferred into the greenhouse, keeping it warmer for longer.

12 (a) (i) It is hotter. Energy is transferred more quickly when something is very hot.

(ii) 120°C

(b) It will be the same rate of transfer, because the temperature is the same.

(c) Yes, it is possible. A pan boils on a cooker, but the temperature stays the same. The energy supplied evaporates the water.

13 (a) $E = Pt$

$\quad = 75 \times 2 \times 60$

$\quad = 9000\,J$

(b) $L = \dfrac{E}{m}$

$= \dfrac{9000}{0.024}$

$= 3.75 \times 10^5 \, \text{J/kg} \approx 3.8 \times 10^5 \, \text{J/kg}$

(c) (ii) If the ice is colder than 0 °C, energy is used to warm the ice up to 0 °C. The specific latent heat is a measure of the energy required to melt the ice at 0 °C, without a temperature change.

(ii) This is a systematic error, because it will consistently give the same incorrect answer.

(d) Clamp the funnel securely to avoid spills. Only turn the heater on when it is inside the ice – it could get dangerously hot.

14 (a) The particles move quickly and randomly in all directions.

(b) The particles move faster as the temperature increases.

15 (a) The fast-moving particles in a gas hit the walls of their container. When the particle hits a wall, it exerts a force. Therefore, there is a pressure.

(b) At a higher temperature the particles move faster, so they hit the walls of their container faster and more often.

16        $P_1 V_1 = P_2 V_2$

$100 \times 0.02 = 500 \times V_2$

$V_2 = 0.02 \times 1/5$

$V_2 = 0.004 \, \text{m}^3$

17        $P_1 V_1 = P_2 V_2$

$700 \times 0.4 = 100 \times V_2$

$V_2 = 2.8 \, \text{m}^3$

Chapter 4 Atomic structure

1 (a) 10^{-10} m

(b) 10 000

2 (a) 8

(b) 16

(c) The 8 positive charges of the protons neutralise the 8 negative charges of the electrons.

3 (a) 5 p 6 n

(b) 16 p 16 n

(c) 64 p 92 n

(d) 93 p 144 n

4 (a) The atomic number is equal to the number of protons.

(b) The mass number is equal to the number of protons and neutrons.

(c) Isotopes are different types of an element; they have the same number of

protons in the nucleus but have different numbers of neutrons – so their masses are different.

(d) An ion is an atom that has gained or lost an electron (or electrons), so an ion is charged.

5 $\dfrac{\text{atomic radius}}{\text{nuclear radius}} = \dfrac{1.8 \times 10^{-10}}{7.5 \times 10^{-15}} = 24\,000$

6 • A beam of alpha particles was directed towards a thin gold foil. Alpha particles are positively charged and energetic.
 • Most alpha particles went through the foil with little or no deviation. A very small number (1 in 10 000) of the alpha particles bounced back.
 • This suggested that the alpha particles were being repelled by a very small, positively charged and massive nucleus. If the nucleus were not massive, the alpha particles would knock it out of the way.

7 The Bohr model of the atom has a small positive nucleus that contains protons and neutrons. Electrons are in orbits around the nucleus. The orbits are circular and there are different energy levels. Electrons can move between the energy levels by absorbing or emitting electromagnetic waves.

8 B and C

9 Radiations emitted from radioactive sources are dangerous to us. The tongs keep the sources away from the teacher's hand, so that fewer particles reach the hand.

10 (a) The activity is the total number of particles emitted per second by a radioactive source.

(b) becquerel

11 (a) A helium nucleus

(b) A fast electron

(c) Electromagnetic radiation

12 No change to either the mass or atomic numbers.

13 (a) $^{241}_{94}\text{Pu} \rightarrow \, ^{237}_{92}\text{U} + \, ^{4}_{2}\text{He}$

(b) $^{237}_{92}\text{U} \rightarrow \, ^{237}_{93}\text{Np} + \, ^{0}_{-1}\text{e}$

(c) $^{59}_{26}\text{Fe} \rightarrow \, ^{59}_{27}\text{Co} + \, ^{0}_{-1}\text{e}$

(d) $^{213}_{84}\text{Po} \rightarrow \, ^{209}_{82}\text{Pb} + \, ^{4}_{2}\text{He}$

(e) $^{32}_{14}\text{Si} \rightarrow \, ^{32}_{15}\text{P} + \, ^{0}_{-1}\text{e}$

(f) $^{229}_{90}\text{Th} \rightarrow \, ^{225}_{88}\text{Ra} + \, ^{4}_{2}\text{He}$

14 activity

15 Throwing a die is a random event. Throwing a six has a probability of 1/6, but we cannot predict with certainty that we will get a 6 on the next throw.

16 (a) C

(b) A

(c) B

17

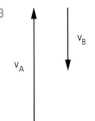

Half-life = 5.2 h

The count rate halves to 6000 after 5.2 h and it halves again to 3000 after 10.4 h (which is 2 half-lives).

[You will get full marks for doing just one half-life.]

18 64 days is 4 half-lives.

So the activity will be $\frac{1}{2} \times \frac{1}{2} \times \frac{1}{2} \times \frac{1}{2}$ of 4×10^5 Bq

$A = \frac{1}{16} \times 4 \times 10^5$ Bq

$= 2.5 \times 10^4$ Bq

19 (a) We are irradiated by radioactive emissions from our environment – rocks, the food we eat and cosmic rays from space. This is background radiation.

(b) Different types of rock contain different radioactive materials. So we get different background counts around the world.

20 (a) A short half-life isotope produces a high activity for a short time. Measurements can be done, then the activity decreases as the isotope decays.

(b) Gamma rays can penetrate the body and be detected outside. Also gamma rays are the least damaging to the body.

21 A patient might go to hospital to receive treatment for a skin cancer: the skin is irradiated by a radioactive source. As soon as the source is removed there is no more radiation – the patient is not radioactive. We can control the radiation.

If there is a leak of radioactive waste from a power station, the ground becomes contaminated and will be radioactive for many years, until all the isotopes have decayed. There is danger for us if we go to a contaminated area. We cannot control contamination.

22 Patient A: this is a relatively low dose so there is little risk. But there are many benefits if a kidney can be cured.

Patient B: this carries quite a high risk and healthy patients should not be exposed to a dose

of 700 mSv. But cancer can kill the patient, so it is worth taking the risk if the radiation destroys the cancer.

23 (a) X has high activity for the first 2 years, so is more dangerous.

(b) After 40 years, X presents little danger, but Y still has a relatively high activity.

24 Neutron

25 Fusion

26 (a) Nuclear fission is the splitting of a large nucleus into two smaller nuclei. This releases energy.

(b) Fission is usually triggered by a neutron being absorbed by a nucleus. Each fission also emits 2 or 3 other neutrons, which can then cause fissions in other nuclei.

27 $^{1}_{0}n + ^{239}_{94}Pu \rightarrow ^{134}_{54}Xe + ^{103}_{40}Zr + 3^{1}_{0}n$

Chapter 5A Forces

1 Vectors: force, velocity, acceleration

2 The direction should be stated too.

3

4 Contact forces (examples): friction, normal contact force, air resistance, tension

Non-contact forces (examples): gravitational, magnetic and electrostatic forces

5 (a) newtons

(b) newtons per kilogram

6 $W = mg$

$= 120 \times 1.6$

$= 192\,N \approx 190\,N$

7 (a) 4 N upwards

(b) 1 N to the left

8 (a)

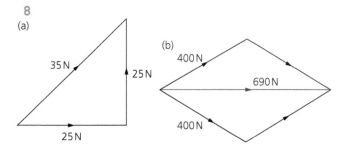

(b)

(a) The force has a vertical and horizontal component each of 25 N.

(b) The resultant force is 690 N.

9 joule, J

10 (a) Energy in your arm's chemical store is being transferred to your arm's thermal store.

(b) No work is done, as the weight is not moved

(c) work = Fs

$= 20 \times 1.3$

$= 26\,\text{J}$

11 (a) work = Fs

$= 30 \times 2.5$

$75\,\text{J}$

(b) When you drag the object, energy is transferred from the chemical store in your body to the thermal store in the surroundings.

12 (a) When something is deformed elastically by forces, it returns to its original shape when the forces are removed.

(b) When something is deformed inelastically, it does not return to its original shape when the forces are moved.

13 (a) 20 mm

(b) $F = ke$

$5 = k \times 0.02$

$k = \dfrac{5}{0.02}$

$= 250\,\text{N/m}$

(c) Work done equals the elastic energy stored.

$E_e = \dfrac{1}{2}ke^2$

$= \dfrac{1}{2} \times 250 \times (0.02)^2$

$= 0.05\,\text{J}$

14 Nm

15 moment = Fd

$= 35 \times 0.80$

$= 28\,\text{N m}$

16 turning moment = force × perpendicular distance

So a longer lever provides a greater turning moment.

17 The moment on the handle is transmitted onto the nail.

$F \times 3 = 45 \times 30$

$F = 450\,\text{N}$

18 The anticlockwise moment balances the clockwise moment.

$450 \times 3.0 = W \times 1.8$

$W = \dfrac{450 \times 3}{1.8}$

$= 750\,\text{N}$

19 (a) moment = Fd

$= 750 \times 0.02$

$= 15\,\text{N m (or 1500 N cm)}$

(b) $15 = F \times 0.3$

$F = 50\,\text{N}$

20 pascal, Pa

21 (a) The molecules in the air are in a constant state of motion. They hit objects, thus exerting a force and therefore a pressure.

(b) There are fewer molecules per m³, so the collisions of the molecules are fewer per second.

22 (a) pressure = $h\rho g$

With greater depth there is more water above to exert a force.

(b) $P = h\rho g$

$1\,013\,000 = 0.76 \times \rho \times 9.8$

$\rho = \dfrac{1\,013\,000}{0.76 \times 9.8}$

$= 13\,600\,\text{kg/m}^3$

23 (a)

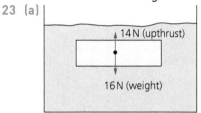

(b) (i) The wood will sink because $W > U$.

(ii) The wood is more dense than water.

Chapter 5B Forces and motion

1 (a) (i) $s = vt$

$100 = v \times 10$

$v = \dfrac{100}{10}$

$= 10\,\text{m/s}$

(ii) $s = vt$

$400 = v \times 46$

$v = \dfrac{400}{46}$

$= 8.7\,\text{m/s}$

(b) Our range for sprinting is 100 m or 200 m. The athletes tire over 400 m.

2 (a) 600 m

(b) Over the part CD. The gradient of the graph is steepest.

(c) 30 s.

(d) $s = vt$

$400 = v \times 40$

$v = \dfrac{400}{40}$

$= 10\,\text{m/s}$

3 10 m/s

4 (a) $s = vt$

$= 55 \times 60$

$= 3300\,\text{m}$

(b) $s = vt$

$11\,000 = 55 \times t$

$t = \dfrac{11000}{55}$

$= 200\,\text{s}$

5 (a) m/s

(b) m/s^2

6 (a) $a = \dfrac{v - u}{t}$

$= \dfrac{5 - 3}{8}$

$= 0.25$ m/s^2

(b) $a = \dfrac{30 - 18}{3}$

$= 4$ m/s^2

or you can write -4 m/s^2, where the minus sign implies deceleration.

7 (a) $a = \dfrac{v - u}{t}$

$= \dfrac{12}{4}$

$= 3$ m/s^2

(b) distance = area under the graph

$= \dfrac{1}{2} \times 12 \times 4$

$= 24$ m

8 (a) (i) $a = \dfrac{v - u}{t}$

$= \dfrac{1}{0.1}$

$= 10$ m/s^2

(ii) $a = \dfrac{v - u}{t}$

$= \dfrac{0.45}{0.1}$

$= 4.5$ m/s^2

(b) The ball reaches a terminal velocity when the drag force equals the weight.

(c) Length of the container equals the area under the graph.

The area is about 4.3 m. So 4.5 is closest.

9 $v^2 - u^2 = 2as$

$50^2 - 30^2 = 2 \times a \times 1000$

$a = \dfrac{2500 - 900}{2000}$

$= 0.8$ m/s^2

10 B

11 (a) (i)

(ii) The book is at rest so the resultant force must be zero.

(b) (i) Equal and opposite forces must be of the same type – e.g. both gravitational, and the equal and opposite forces (in Newton's third law) act on different bodies.

(ii) The gravitational pull of the book on the Earth.

12 acceleration $= \dfrac{\text{force}}{\text{mass}}$

So by having a low mass the car can accelerate and decelerate rapidly.

13 (a) $a = \dfrac{v - u}{t}$

$= \dfrac{18 - 10}{12}$

$= 0.67$ m/s^2

(b) $F = ma$

$= 1500 \times 0.67$

$= 1000$ N

14 The balloon pushes air backwards out of the balloon. Therefore, the air exerts an equal and opposite force on the balloon forwards.

15 (a) The resultant force on the boat is zero.

(b) $a = \dfrac{v - u}{t}$

$= \dfrac{20 - 11}{6}$

$= 1.5$ m/s^2

(c) (i) $F = ma$

$= 2500 \times 1.5$

$= 3750$ N ≈ 3800 N

(ii) Assuming the drag remains at 8000 N

$F = 8000 - 3750$

$= 5250$ N ≈ 5300 N

(iii) The boat goes at a constant speed, so drag $= 5250$ N ≈ 5300 N.

16 A drunk driver

17 A muddy road; the speed of the car

18 (a) $s = vt$

$= 15 \times 0.4$

$= 6$ m

(b) $s = vt$

$= 15 \times 0.6$

$= 9$ m

19 kg m/s

20 momentum $= mv$

$= 1250 \times 20$

$= 25\,000$ kg m/s

21 (a) force $= \dfrac{\text{change of momentum}}{\text{time}}$

By wearing a seat belt, a passenger makes sure they use as much time as possible to stop when a car brakes suddenly. Without a seat belt a passenger stops in a short time and the force on them is large.

(b) A helmet makes the time of impact on a head longer (as there is padding inside). Thus the force is less. Also the helmet protects the head against sharp objects, where the pressure would be large.

22 (a) $F = \dfrac{\text{change of momentum}}{\text{time}}$

$= \dfrac{0.16 \times 30}{0.1}$

$= 48\,\text{N}$

(b) This makes the time of stopping longer and reduces the force.

23 momentum before the collision = momentum after the collision.

$1.5 \times 0.6 = (1.5 + 1.0)v$

$v = \dfrac{1.5 \times 0.6}{2.5}$

$= 0.36\,\text{m/s}$

Chapter 6 Waves

1 (a) (i) See Figure 6.2

(ii) See Figure 6.3

(b) (i) Water wave, electromagnetic wave

(ii) Sound wave, p-wave (seismic wave)

2 Energy – you can feel a pulse at the other end of the spring.

Information – you can invent a code or use Morse code to send information.

3 The ball goes up then down. It does not go from side to side.

4 A has a higher amplitude and a higher frequency than B.

5 (a) (i) $f = \dfrac{1}{T}$

$= \dfrac{1}{0.02}$

$= 50\,\text{Hz}$

(ii) $f = \dfrac{1}{T}$

$= \dfrac{1}{0.001}$

$= 1000\,\text{Hz}$

(b) (i) $T = \dfrac{1}{f}$

$= 10^{-9}\,\text{s}$

(ii) $T = \dfrac{1}{f}$

$= \dfrac{1}{2 \times 10^6}$

$= 5 \times 10^{-7}\,\text{s}$

6 $v = f\lambda$

$1.2 = f \times 0.2$

$f = 6\,\text{Hz}$

7 (a) Transverse

(b) (i) 40 cm

(ii) 2.0 m

(c) (i) 2 Hz

(ii) 0.5 s

(d) $v = f\lambda$

$= 2 \times 2$

$= 4\,\text{m/s}$

8 (a) Refer to Figure 6.8 and the associated text.

(b) Refer to Figures 6.10 and 6.11.

9 (a) The frequency is unchanged.

(b) The wavelength becomes shorter in carbon dioxide.

10 Refer to Figure 6.14.

11

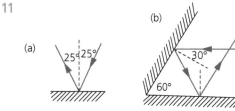

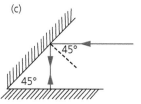

12 (a) A wave is absorbed when it enters a medium but does not travel through it. The wave energy is absorbed by the medium.

If a wave passes through a medium it is transmitted.

(b) (i) Rock reflects sound from air.

(ii) Air

(iii) Soft foam rubber

13 (a) Sound with a frequency higher than 20 kHz

(b) (i) Forming images of unborn babies

(ii) Depth finding

14 $v = f\lambda$

$1600 = 4 \times 10^6 \times \lambda$

$\lambda = \dfrac{1600}{4 \times 10^6}$

$= 0.0004\,\text{m or } 0.4\,\text{mm}$

15 (a) P-waves are longitudinal waves; s-waves are transverse waves.

(b) S-waves are reflected off the core; p-waves are transmitted by the liquid core.

16 (a) $v = f\lambda$

$1500 = 25 \times 10^3 \times \lambda$

$\lambda = \dfrac{1500}{25 \times 10^3}$

$= 0.06\,\text{m}$

(b) (i) $s = vt$

$= 1500 \times 0.18$

$= 270\,\text{m}$

So the submarine is 135 m deep. (The waves travel to the submarine and back.)

(ii) $s = 1500 \times 0.24$

$\quad = 360\,m$

So the submarine is 180 m deep.

It has travelled 180 − 135 = 45 m in 30 s.

$v = \dfrac{s}{t}$

$\quad = \dfrac{45}{30}$

$\quad = 1.5\,m/s$

(c) The rubber absorbs the ultrasound, so that there is no echo for the ship to detect.

17 (a) $\quad v = f\lambda$

$\quad 3 \times 10^{8} = f \times 6.5 \times 10^{-7}$

$\quad f = \dfrac{3 \times 10^{8}}{6.5 \times 10^{-7}}$

$\quad\quad = 4.6 \times 10^{14}\,Hz$

(b) $\quad v = f\lambda$

$\quad 3 \times 10^{8} = 2 \times 10^{6} \times \lambda$

$\quad \lambda = \dfrac{3 \times 10^{8}}{2 \times 10^{6}}$

$\quad\quad = 150\,m$

18

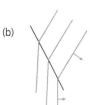

(a)　　　　(b)

19

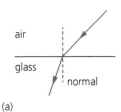

(a)　　　　(b)

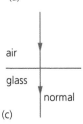

(c)

20 The p-waves refract towards the normal when they travel from the mantle into the outer core. So they travel more slowly in the outer core than in the mantle.

21 (a) Ultraviolet, x-rays, gamma rays

(b) • Ultraviolet – aging skin and skin cancer
   • X-rays and gamma rays can cause mutations to DNA and then cancer

22 (a) Microwaves – satellite communications, and cooking

(b) Infrared – heating, night photography

23 Radio waves, microwaves, infrared, light, ultraviolet, x-rays, gamma rays

24 • Use the same amount of wax for each marble.
   • Make sure the marbles are the same mass.
   • Use metal sheets of the same thickness.
   • Keep the heater equal distances from each sheet.

25 (a) and (b) produce the same list: dull black, shiny black, dull white, shiny metallic.

26 (a) Rays converge at a real image, so it can be projected on to a screen.

   A virtual image only appears to be there, because rays appear to diverge from the image.

(b) Concave

27 (a)

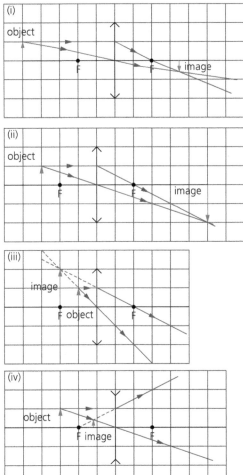

(b) (i) and (ii) are real; (iii) and (iv) are virtual

(c) (i) 0.6

(ii) 2

(iii) 2

(iv) 0.4

28 (a) Image must be 12 cm from the lens.

(b)

(c) $F = 4$ cm

29 (a) Black

(b) Green

(c) Black

30 Trousers – black; shirt – red; cap – black

31 A red object looks red in white light because it reflects red and absorbs all the other colours. Each colour of light has a different wavelength.

32 (a) $v = f\lambda$

$3 \times 10^8 = f \times 4.8 \times 10^{-7}$

$f = \dfrac{3 \times 10^8}{4.8 \times 10^{-7}}$

$= 6.25 \times 10^{14}$ Hz

(b) $v = f\lambda$

$3 \times 10^8 = 5 \times 10^{14} \times \lambda$

$\lambda = \dfrac{3 \times 10^8}{5 \times 10^{14}}$

$= 6 \times 10^{-7}$ m

33 (a) More infrared radiation is absorbed than is emitted.

(b) The meat absorbs and emits the same amount of infrared radiation.

(c) More infrared radiation is emitted than is absorbed.

34 A perfect black body absorbs all the electromagnetic radiation that falls on it. It is also the best possible emitter of radiation.

Chapter 7 Magnetism

1 Iron, steel, cobalt, nickel

2 Two. Like poles repel; unlike poles attract.

3 (a) A permanent magnet retains its magnetism.

(b) A permanent magnet can be repelled by another permanent magnet. An induced magnet only becomes a magnet when it is in the magnetic field of another magnet.

4 See Figure 7.4

5 A south pole

6 (a) The weight of the clips

(b) (i) Induced magnets

(ii) The top of each clip is a south pole, and the bottom of each clip is a north pole.

(c) They are induced magnets. So the top of each clip would then be magnetised north, and the bottom magnetised south.

7 The head of each pin is magnetised south, so they repel.

8 • Increasing the current.

• Adding more turns of wire.

• Pushing the coils closer together.

• Putting an iron core into the solenoid.

9 Refer to Figure 7.8.

10 The thumb points in the direction of the current; the fingers show the direction of the field lines. Refer to Figure 7.9.

11 Refer to Figure 7.10.

12 (a) 1, 2, 4 point left; 3, 5, 6 point right.

(b) The right-hand end

(c) All the compasses reverse direction.

13 (a) Down

(b) Down

(c) Up

(c) Up

14 Increasing the current; using magnets with a stronger field.

15 tesla, T

16 (a) $F = BIL$

$= 0.18 \times 1.3 \times 0.05$

$= 0.01$ N

(b) Zero

17 The wire exerts a force on the two magnets.

18 (a) Zero. BC is parallel to the field.

(b) The forces lie along the same line, so will not turn the coil.

19 (a) The strength of the magnets; the size of the current; the number of turns of wire.

(b) The split-ring commutator switches the direction of current in the coil, each half turn. This makes sure the turning forces are always in the same direction.

20 (a) Refer to Figure 7.20.

(b) When the current flows one way, the cone is pushed out; when the current reverses in direction, the cone is pushed in. So with an a.c. supply the cone vibrates backwards and forwards.

21 (a) Moving the wire faster; using a stronger magnetic field.

(b) Reversing the direction of movement; reversing the magnetic field.

22 (a) Zero

(b) In each case the meter deflects to the right.

(c) Move the magnet faster; use a magnet with a stronger magnetic field.

23 (a) Vertical at times A, C, E. Horizontal at times B and D.

(b)

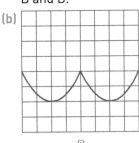 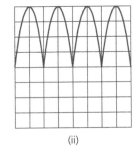

(i)   (ii)

24 Sound waves cause a diaphragm in the microphone to vibrate backwards and forwards. The diaphragm is attached to a small coil that oscillates close to a magnet. The changing magnetic field induces an alternating potential difference.

25 (a) The output p.d. of a step-up transformer is greater than the input p.d.

(b) $n_s > n_p$

(c) A charger for a phone, shaver, toothbrush

26 Line 1: 30 V step–down

Line 2: 250 V step–up

Line 3: 1200 turns step–down

Line 4: 120 turns step–down

27 (a) A p.d. is only induced in the secondary coil when there is a changing magnetic field. Direct current produces a constant field, so no p.d. is induced. Alternating current induces a p.d. due to the changing field.

(b) (i) When the p.d. is stepped up on the power lines, the current flowing decreases. This reduces the power dissipated in the lines. $P = I^2R$; so P is less for a smaller current.

(ii) The p.d. must be reduced to 230 V for safety reasons.

28 $$\frac{V_s}{V_p} = \frac{n_s}{n_p}$$

$$\frac{V_s}{110} = \frac{20}{400}$$

$$V_s = \frac{110}{20}$$

$$= 5.5\,\text{V}$$

$$I_s = \frac{V}{R}$$

$$= \frac{5.5}{125}$$

$$= 0.044\,\text{A} \approx 0.04\,\text{A}$$

Chapter 8 Space physics

1 The Solar System is the Sun plus planets, moons and other bodies. Solar implies belonging to the Sun – all objects in the Solar System are under the influence of the Sun's gravitational pull.

2 A dwarf planet orbits the Sun; a moon orbits a planet (which orbits the Sun).

3 The Milky Way

4 B, E, F, C, D, A

5 (a) Nuclear fusion

(b) (i) Main sequence

(ii) A black dwarf

6 (a) A main sequence star is in a stable state, fusing hydrogen nuclei to form helium nuclei.

(b) Refer to Figure 8.2.

The outward pressure caused by the thermal energy of nuclear fusion, balances the force of gravity which tends to collapse the star.

7 The galaxy contains large clouds of hydrogen gas and dust. Gravity acts over infinite distances and pulls these clouds together. When the clouds collapse, their store of gravitational potential energy is transferred to kinetic energy and then thermal energy; a hot protostar is formed. The temperature rises and nuclear fusion begins, and a main sequence star, fusing hydrogen, is born. After billions of years the hydrogen runs out. The star collapses and the core becomes even hotter. Now the fusion of helium starts and the star becomes a red giant. When the helium runs out, the star collapses to a white dwarf, and eventually a black dwarf when it cools.

8 (a) The pull of Earth's gravity

(b) A satellite moves in a circular orbit at a constant speed.

(Some satellites move in elliptical orbits, but we do not need to describe their motion.)

9 (a) $s = vt$

$$5 \times 10^7 = v \times 2 \times 60 \times 60$$

$$v = \frac{5 \times 10^7}{2 \times 60 \times 60}$$

$$= 6980\,\text{m/s} \approx 7000\,\text{m/s}$$

10 (a) Speed is scalar, so just has a magnitude. Velocity is a vector, so we must define a direction.

(b) A moon's speed is constant, but its velocity is always changing, because its direction of travel changes. (See Figure 8.4)

11 She can deduce that the galaxy is moving away from us.

12 True: B, E; false: A, C, D

13 The Universe is populated by billions of galaxies. Wherever we look, in all directions, galaxies show a red-shift in their light. This tells us galaxies are moving away from us. The more distant galaxies have a greater red-shift and are therefore moving faster. We can deduce from this that all the galaxies were once in the same place. Thus it seems likely that matter was thrown outwards with a great explosion a long time ago.